T0130865

This book was originally published as *The future of food. Een nieuw recept voor de voedingssector*, LannooCampus, 2021.

D/2021/45/611 – ISBN 978 94 014 8076 5 – NUR 946, 800

Cover design: Ben Meulemans
Interior design: Gert Degrande | De Witlofcompagnie
Translation: Ian Connerty
With editorial contributions by Pauline Neerman

LannooCampus Publishers is a subsidiary of Lannoo Publishers, the book and multimedia division of Lannoo Publishers nv.

LannooCampus Publishers
Vaartkom 41 box 01.02
3000 Leuven
Belgium
www.lannoocampus.com

P.O. Box 23202
1100 DS Amsterdam
Netherlands

JORG SNOECK AND
STEFAN VAN ROMPAEY

THE FUTURE OF FOOD

A new recipe for the food sector

Lannoo
Campus

// CONTENTS //

FOREWORD

Let's eat!

What did you have for breakfast this morning? Perhaps it was a low-carb granola with berries and skyr, washed down with an oatmeal drink – or, if you are a fan of green tea, with a chai latte. Maybe for lunch you will tuck in to a tasty avocado salad with quinoa and pomegranate seeds. Or do you prefer a vegetable burger? Just twenty years ago, none of this would have been thinkable. Food is something personal. Often, it is a family story, with recipes and customs being passed down from generation to generation. But not any more. Eating habits rooted in tradition are now changing at lightning speed.

Nowadays, we regard it as normal that we have instant access to the most delicious food products from the four corners of the world. However, we are gradually becoming aware that this process of globalisation also has a darker side. Food comes at a price. It has an impact on our health, on the environment and on society as a whole. What will happen when we have ten billion mouths to feed in 2050? Food production will need to double, but we only have one planet of limited size. This means that we will need to do things differently – and better. Technology can help, but what we really need is a structural transformation. The entire food system must be redesigned.

The farmers in the South who grow the cacao for our delicious Belgian chocolate are unable to earn a living wage from their labours. The farmers closer to home who invested in pig farming have seen the price for their meat rise and fall like a roller-coaster as a result of international market mechanisms over which they have no control. Global warming is endangering coffee production. A fungal infection might clear the shelves in Western supermarkets of people's favourite variety of banana. A strong increase in demand from growing markets – not least from China – risks knocking world trade out of balance and threatens to create shortages of some natural resources and foodstuffs.

Supermarket chains publish fine-sounding sustainability reports, but at the same time continue with promotional stunts for products that are neither sustainable nor healthy, seeing this as the only way to remain competitive in a fierce market. In the Western world, obesity is now a more serious problem than hunger. The food industry is unintentionally making us ill, with its development of a huge range of super-sophisticated, near irresistible but seriously over-processed goodies at knock-down prices. As a result, a counter-movement is now gaining ground that hopes to bring more organic, vegetable, gluten-free, lactose-free, allergy-free, sugar-free and clean-label products to supermarket shelves.

The corona crisis has moved all these changes into overdrive. Consumers have become more demanding than ever. They increasingly distrust multinational food and retail conglomerates. They want guarantees about the safety and quality of their food. They insist on transparency about the origin of ingredients and the working conditions in countries where food products are made. They expect faultless service and seamless ease of use. Ideally, they want their groceries delivered at home in fifteen minutes.

The corona crisis has also taken the digitalisation of our world to the next level. This has had huge consequences for the food sector. Online shopping for groceries has finally made a major breakthrough. The chances of things returning to the way they once were are zero. New digital distribution models now pose a serious threat to the traditional supermarket channel: it is now much easier for brands and start-ups to approach consumers directly, with no need for intermediaries.

Another change that is here to stay is the way in which consumers have organised their lives since the outbreak of corona. Health and safety are now more prominently on the agenda than ever before. This is something that manufacturers and retailers will need to take into account. Similarly, the fact that many more people will continue working from home – even after corona – will have a lasting impact on eating and shopping habits. The traffic at traditional commuter locations such as city centres, industrial estates and train stations will fall significantly, with unwelcome consequences for the turnover of the shops and catering outlets at these locations. New eating and purchasing moments will increasingly be transferred to the home or the local neighbourhood. The frequency of visits to physical stores will decline still further.

More than ever before, there is a growing awareness that everyone is connected with everyone else in a globalised food ecosystem. Notwithstanding greater interest in short chains and local production, consumers and retailers will not be willing to give up the benefits that globalisation has brought them. However, a degree of adjustment is certainly necessary. During the pandemic, we have all experienced the vulnerability of the global distribution networks at first hand. Moreover, the relegation of the climate crisis to the background during the corona period was purely temporary. It will soon re-emerge centre stage, and with a renewed sense of urgency. We cannot continue to exhaust the planet as we have done in the past. It is not possible to maintain a resilient food system unless all the partners in the ecosystem receive a fair share of the resultant produce – not only to allow them to live a dignified human existence, but also to give them the opportunity to invest in a sustainable food future.

In the decades ahead, it is crucial that all the links in the food chain respond quickly and flexibly to meet these challenges. We need organisations that are capable of mobilising worldwide networks; of inspiring younger generations; above all, of thinking outside the box. New business models will throw down the gauntlet to established market leaders. Digital natives will look at market relationships through Google glasses. Technology will be a game changer for the entire chain, from farm to plate. Brands and service models will no longer require a physical footprint to make their presence felt. The interaction between producer, retailer and consumer will be shifted to the cloud.

The content of our diet will also be very different in twenty years' time. Will we still be eating animals? Or will seaweed be the new protein of the future? Will we still cook our food? Or will we leave that to 3D printers, which will develop personalised recipes based on our DNA and our microbiome? Will be willing to swallow – both literally and metaphorically – ingredients that are prescribed by machines to support not only our physical but also our mental health? Will we still do our daily and weekly shopping? Or will this be delegated to our fridge, smart algorithms and self-driving delivery vehicles?

One thing that almost certainly will not change is the connective power of food: the stories that we tell while consuming our daily meals around the family table, the recipes we exchange with friends, the experiences we share over a glass of good wine, the chocolates that we give to each other to celebrate special occasions ...

In this book, we will take you on an exciting journey, a voyage of discovery through the food system of tomorrow. Enjoy!

ACKNOWLEDGEMENTS

Writing a long and richly filled book is a huge undertaking. But you don't really know just how huge until you start! We have several months of intense effort behind us, but the enthusiastic responses of the first proofreaders have left us feeling proud, pleased and even somewhat emotional. We have been fortunate (and are very grateful) that a number of respected top managers from the food sector have been willing to read our book and comment on it critically. In particular, we would like to thank Frans Muller (Ahold Delhaize), Jef Colruyt (Colruyt Group), Koen Slippens (Sligro Food Group), Nils van Dam (Milcobel), Hein Deprez (Greenyard), Wim Destoop (PepsiCo), Dirk Van den Berghe (ex-Walmart) and Lieven Vanlommel (Foodmaker) for their constructive suggestions!

We also owe a massive debt of gratitude to all our colleagues and partners at RetailDetail: we could not possibly have completed this book without their help. Together, we have built up a network and knowledge platform, the mission of which is to enlarge the collective brain of the retail sector. This knowledge is the result of intense collaboration, dialogue, discussion and interaction. Teamwork makes the dream work! A special word of thanks must also go to Pauline Neerman, the co-author of the prize-winning management book *The Future of Shopping*, for her valuable contribution to our book. We truly have a top team!

Jorg & Stefan

Major challenges on the menu

The food system under pressure

Welcome to the anthropocene. We live in the age of humans, the animal species that has firmly established itself at the very top of the survival ladder. It has to be said: we have done very well for ourselves. We control and rule this planet. Thanks to the industrialisation of agriculture, the production of food has never been as efficient as it is today. Safe and high-quality food is available all around the world on a scale never previously seen, both in terms of price and ease of consumption.

Even so, you get the sense that something somewhere is not quite right. The fact that human beings are now lord and master of the planet has had a number of unforeseen consequences. Disease and pandemics? Thanks to a few bats in China, the entire population of the world has had to stay a metre and a half away from one another for more than a year. Natural disasters? The emissions from factories are melting the icecaps and people are being drowned on Thai beaches. The world is facing some serious global challenges that can no longer be denied.

What has all this got to do with the daily purchases that we make in our local supermarkets? Quite a lot. In the near future, we will be using more of everything than our planet can sustain. The critical point of no return will be reached roughly halfway through this present century; the point when it will no longer be possible for nature to recover. More than five hundred species of land animal are on the point of extinction as a consequence of human activity. Most will probably disappear for ever within the next twenty years, more than disappeared in total during the preceding hundred years. Scientists refer to this as the sixth wave of extinction and it is (so far, at least) the tragic culmination of the anthropocene era – the age in which human beings determine the fate of the Earth. And the fifth wave of extinction? That was when the once omnipotent dinosaurs died out ...

Food is the strongest lever

In a relatively short period of time, the way in which we organise food production and agriculture has become a key determining factor for the future of our planet. The loss of genetic diversity is making itself felt in the food system. The amounts of nitrogen, phosphorus and carbon already exceed the limit of what the planet can support and the irreversible decline of biodiversity threatens to drastically change the biosphere. Equally drastic changes in the climate are causing new diseases (as we have discovered on an unparalleled scale with COVID-19), poverty, natural disasters and conflicts. Expenditure on health care is now one of the most significant economic burdens in both developed and developing countries, and was so even before the corona pandemic.

There are now so many of us, and so many more are on the way, that our planet is coming under increasing pressure. Meeting our essential basic needs is becoming an ever greater challenge. In short, how can we keep feeding the world? Food and the food system form the link between health and ecological sustainability. Both are crucial for the future of our species. It is therefore something of a paradox that food production currently poses the most serious threat to the environment. Today's food production accounts for no less than a quarter of all human greenhouse gas emissions. Each year, the conversion of uncultivated land into agricultural land is responsible for one-sixth of these emissions, resulting in the eradication of more animal and plant species than those lost through climate change (Funabashi, 2018).

'Food is the single strongest lever to optimise human health and environmental sustainability on Earth. However, food is currently threatening both people and planet.' This was the stark warning contained in the EAT-Lancet report (2019), complied by 37 leading scientists from 16 different countries. To keep on feeding an ever-growing world population, we not only need more food; we also need healthier and more sustainable food.

The trilemma of the 21st century

The realisation that these challenges are inextricably connected with each other has meant in recent years that food has risen to the top of the agenda, both for companies and for politicians. Protein strategies (Flanders), national food strategies (England) and 'farm-to-fork' plans (EU) now follow each other in quick succession, while the business world is equally convinced of the urgent need for a 'food transition' (Carrefour).

Already in 2017, Wouter Kolk, a top executive at Ahold Delhaize, warned of the risks of food shortages caused by the increase in world population and the changes in demand patterns in developing countries. 'The Chinese are now knocking at the door of our mandarin supplier in Spain. They also want to buy his harvest. So where will I get my mandarins from in future?' (Rijlaar, 2017). The first tangible effect of this increased competition: rising prices.

In its 'Act for Food' programme, the French retailer Carrefour identified four key pillars that need to be addressed for a successful food transition: food waste, the full cost price of food, soil-conserving agriculture, and new trade and retail models. The changing lifestyle and standards of the worldwide consumer are setting new demands in terms of how, when and in what form their food reaches them.

We are facing a global trilemma between food, the environment and health, a trilemma in which it currently seems that one of the three elements needs to suffer for the benefit of the other two. If we wish to solve this trilemma and achieve a sustainable solution, a drastic and fundamental reform of our food system is inevitable, and this along the entire value chain: from production through distribution to consumption.

Producing for tomorrow

Indoor versus outdoor cultivation

If we carry on as we are, we will soon exhaust nature – both literally and figuratively – and create more and more desert: dry and infertile land from which all life has been drained. Solutions to continue agricultural production without the need for additional agricultural land, with all the negative consequences this implies, are looking increasingly at the possibilities of vertical farming (in vivo) and the production of cultured meat in laboratories (in vitro). In the long run, we could even grow our salad in this way.

Even so, this 'indoor cultivation' cannot fully answer all our food and climate challenges. Our existing ecosystems also urgently need help to bring the current decline in the ecological condition of our planet, with all the associated health risks this involves, to a halt. These high-tech food innovations are still only in the test phase, so that our knowledge is both limited and fragmented. For example, we do not yet know enough about new pathogens, such as allergies and nerve diseases, or what the impact of these new foods and cultivation methods is likely to be.

Moreover, at the present time 77% of basic products and foodstuffs is produced by small or medium-sized agricultural enterprises. Some 87% of all agricultural land is still cultivated by small-scale farmers working on family-based farms. You don't need to be a genius to realise that this type of farming cannot be expected to immediately make the switch to capital-intensive and high-tech vertical greenhouse facilities with hydroculture.

If we wish to feed more than 9 billion mouths by 2050, a large proportion of the food will still need to come from these small-scale farmers. For this reason, a combined approach will be necessary. Fortunately, experiments in Japan have shown that it is possible with new methods to grow a wide variety of crops on as little as 3,000m² of land, producing results for which traditional agriculture would require an entire region. This 3,000 m² would be enough to maintain the natural balance of the region in which it is situated.

From 30 to 30,000 edible crops

The agricultural industry has undergone a major transformation in recent years. Thanks to modern technology, agricultural yields have increased exponentially. Better weed suppression, crop control and the cultivation of new and more productive varieties have resulted in record harvests that meet the strictest quality expectations. New technologies, such as artificial intelligence, drones and data-driven ICT systems, have created high-tech farms, where 'smart farming' allows precision agriculture to be implemented. In fact, artificial intelligence is already capable of running a farm on its own.

For farmers who want to stay in the game, some degree of scaling-up seems inevitable; all the more so when bearing in mind the current miniscule margins that result from worldwide competition. Mega-farms first appeared in the US, but are now common everywhere. At least 72% of all poultry and 55% of pork is produced in this kind of factory farm (Harvey e.a., 2017). Their huge economies of scale not only dramatically cut production costs, but also force their smaller rivals out of business.

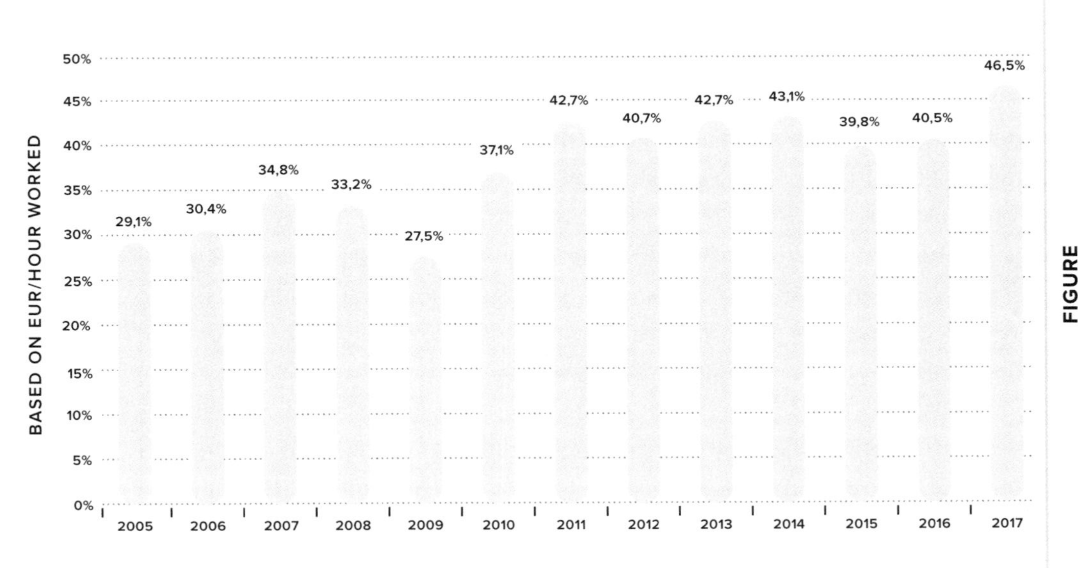

FIGURE
Entrepreneurial income per family work unit compared to average wages in the economy, EU-28

Source: EU DG Agriculture and Rural Development, Agricultural and Farm Income, p.15 (2018).

In 2017, the family income of farmers amounted to less than half the average wage. During the economic crisis of 2009 it was even worse: they earned only a quarter of what they could otherwise have earned outside agriculture, which was a clear sign of the sector's volatility and sensitivity to economic fluctuations. Similarly, the wages of agricultural labourers also amount to less than half of what other kinds of labourers are paid. These low incomes explain why in Europe fewer and fewer agricultural enterprises are able to survive and also why reform measures in favour of health and the environment meet with resistance from the sector.

Of course, none of this changes the heart of the matter: the farmers of the future will need to make significant progress if they wish to feed all the world's hungry mouths. Roughly 85% of the necessary growth in world production during the next ten years must be achieved through increased yields resulting from the more intensive use of the means of production, greater investment in production technology, and improved cultivation practices (OECD & FAO, 2020). More intensive use of the land by harvesting at least twice per year can achieve another 10% of the target growth, while the remaining 5% will need to be accounted for by an increase in the area under cultivation (a much smaller proportion than in recent decades).

Unfortunately, this creates a kind of vicious circle: money for the necessary investments can only be found through scaling up, but this inevitably puts the climate and the ecosystem under even more pressure. In the Netherlands, for example, an attempt to improve air quality means that the government authorities are no longer willing to grant planning permission to enterprises if their expansion leads to an increase in nitrogen emissions. However, this in turn means that agricultural enterprises, who are responsible for some 75% of nitrogen deposition, feel especially targeted, much to their displeasure. Will agriculture ever be able to escape from this downwards spiral? Is there a way to make the farming profession seem more attractive and more profitable? Time will tell.

Greater productivity also has a downside. In 2019, a fungal infection threatened to destroy tens of thousands of hectares of banana plantations, with potentially devastating consequences for supply and prices. Because only one type of banana is grown for the Western market – the sweet Cavendish variety – all the bananas were susceptible to the infection, so that the results were soon catastrophic.

For this reason, some scientists argue that to rely primarily on increasing the scale and the efficiency of today's agriculture will not be sufficient to produce the sustainable solutions we need, not only in terms of feeding the world's growing population but also in terms of combating climate change. The bad news is that this kind of monoculture is currently the most dominant characteristic of our entire food system: worldwide, we derive 90% of our nutritional calories from just thirty different varieties of crop, whereas historically there are more than thirty thousand! But that is also good news: it means that there are still many unused sources of food waiting to be tapped.

Future-proof distribution

Failing global supply chains

The recent corona pandemic made painfully clear that the current food chain is far too long. When people began stockpiling at the start of the crisis, customers were soon confronted by row after row of empty shelves. The global supply chain could not respond quickly enough to the rapid increase in demand. The situation was made even worse by faltering logistics, closed borders and empty factories, whose workers were all infected.

The Dutch supermarket group Albert Heijn is an illustration of how the pandemic helped to redefine the relationship and the relative balance of power in the food supply chain: the retailer is now able to impose a penalty on its suppliers if they fail to make their agreed deliveries – even in the event of 'epidemics and pandemics'. This is now the new normal: producers can no longer hide behind the excuse of *force majeur* or 'act of God'. The chain has learnt its lesson and wishes to ensure that 'customers can always find what they want' (RetailDetail, 2021).

A combination of factors – increasing production costs in developing countries, international trade disputes, sustainability initiatives, etc. – meant that even pre-COVID a growing number of enterprises were looking to move production closer to their home country. The pandemic has simply helped to speed up this switch to local suppliers. A short, flexible and resistant supply chain looks set to become the rule rather than the exception in the years ahead.

The shift to the domestic economy

Consumers are also showing more and more interest in local food systems: they associate 'local' with familiar, ecological, fresh and high quality. But will prices stay the same? And what will happen if they don't? Research suggests that financially insecure consumers, who are likely to be one of the long-term outcomes of the corona crisis, will become increasingly price conscious and value driven (PMA, 2020).

Financial concerns and uncertainties will make consumers cautious when it comes to their purchasing decisions, while at the same time the greater focus on the domestic economy will result in a changed pattern of expenditure. During the pandemic, eating at home became a worldwide norm and it is a habit that many consumers intend to keep once the crisis has passed. Results from research carried out by the Royal Bank of Canada revealed that 66% of Americans and 53% of Canadians prefer home-cooked meals and the majority in both countries said that in future they would only spend the same or even less on 'eating out'. If this trend is confirmed, it will mean a permanent shrinkage of the out-of-home distribution channel and a shift in market share towards the retail channels.

Within the domestic economy there has already been a considerable increase in the number of home-delivered meals. In 2018, these types of meals – think of Takeaway.com and Deliveroo – were good for a turnover of 23.5 billion dollars, but this figure is expected to increase to 99.7 billion dollars by 2027, which is equivalent to an annual growth of 17.4%. In particular, it is the Asian-Pacific countries that are leading the way, mainly as a consequence of changed eating habits among the young and the increase in e-commerce in developing countries (Stratistics, 2020).

The focus on the domestic economy will also mean that buying your daily/weekly groceries online will become a permanent feature in purchasing patterns. The global market for this type of purchase is estimated to increase from 189.81 billion dollars in 2019 to 1.1 trillion dollars in 2027. The 'fresh products' category alone is set to grow by more than 22% each year (Grand View Research, 2020).

The hollowing out of the food distribution network

What will all this mean for the logistical, ecological and economic costs of e-commerce? Perhaps surprisingly, the German logistical player Hermes decided to close down its Liefery 'fast-and-fresh' delivery service at the end of 2020. Their explanation: 'Although there has been increased demand, especially this year, the margins for Hermes as a pure logistic service provider in this low-priced product segment are so small that there was no prospect of making a profit, not even in the medium-long term' (Penrose, 2020).

Now that customers have learned how to spend less time in shops and supermarkets, the challenge for the food retailers is to develop new models and techniques to supply their needs. How do you let consumers become acquainted with your new products? How can you stimulate impulse buys in a digital environment? How can traders meet the demand for a short, local chain? Fast-growing food-as-a-service models and direct-to-consumer sales are hollowing out the role of the traditional food distribution network and will continue to do so.

FOOD LOSS AND FOOD WASTE

If we wish to continue feeding the growing world population with the produce derived from the limited space on our planet – today, 38.6% of the earth's available land surface is already used for agricultural purposes and further deforestation should be avoided at all costs – consumers will need to look at what they can find on their own plates and how they deal with it. It is already hard enough to provide enough food for people to eat, without a significant part of it ending up on the rubbish dump.

One-third of all food is lost or wasted. This is equivalent to 1.3 billion tons per annum. It is further estimated that 25% of all nutritional calories and up to half of the total food weight in the world is never consumed. The mountain of food that ends up in dustbins in Europe alone would be enough to feed 200 million people. Research suggests that in Great Britain 61% of food wastage – some 4 million tons each year – could be avoided with better management and control.

For this reason, reducing the amount of food loss and/or waste is an important part of most scientific scenarios for the global food strategy of the future. The food and agricultural organisation (FAO) of the United Nations has set the ambitious target of cutting food wastage per capita by half by 2030 (in comparison with 2016).

It is important to make a distinction between food loss and food waste. In official terms, food waste refers to food that is of good quality and fit for consumption, but does not get consumed because it is discarded, either by retail outlets or by consumers: in private homes, restaurants, supermarkets, etc. In short, it is eatable food that is thrown away: the part of your meal you don't eat, products that have passed their sell-by date or are otherwise unsellable, etc. Inevitably, this kind of waste occurs most frequently in rich countries, although the precise scale is not known because of a lack of reliable data. The UN is developing indicators that will allow this to be assessed more accurately and will also help to propose the most effective counter-measures.

In poorer countries, a lot of food fails to make it as far as the marketplace, largely as a result of unreliable transport and storage facilities. This is what we regard as food loss and it covers everything from the moment of harvesting until the food arrives at the retail level. Worldwide, 13.8% of all food is lost during this stage: on the farm, in transport, during processing, packing and preserving, in wholesale outlets. This loss is worth an estimated 400 billion dollars per annum (FAO, 2020). It is worth noting that (perhaps unexpectedly) Europe scores badly in this respect: although Central and Southern Asia lose roughly 20%, Europe's 15.7% is the second worst score for losses *en route* to the market. The comparable figure for Australia is just 5.8%.

Consumption in the future

Two billion more people, 50% more food

The world's population of wild animals has been halved since 1970. During the same period, its population of human beings has doubled. In just six generations, human population has exploded. 10,000 years ago, there were roughly 5 million people on Earth, a total that grew slowly until around 1800. By the start of the 19th century, the number had risen to 1 billion. By 1960, the landmark of 3 billion had been passed, before skyrocketing to 7 billion in 2011. By 2050, it is expected that the world's population will be somewhere in the region of 9.7 billion, but a corresponding rise in general prosperity means that the demand for food will increase by more than half (Deloitte, 2020). This corresponds with the prediction of the United Nations made in 2009 that world food production would need to be doubled by 2050, if everyone on the planet was to be kept properly fed.

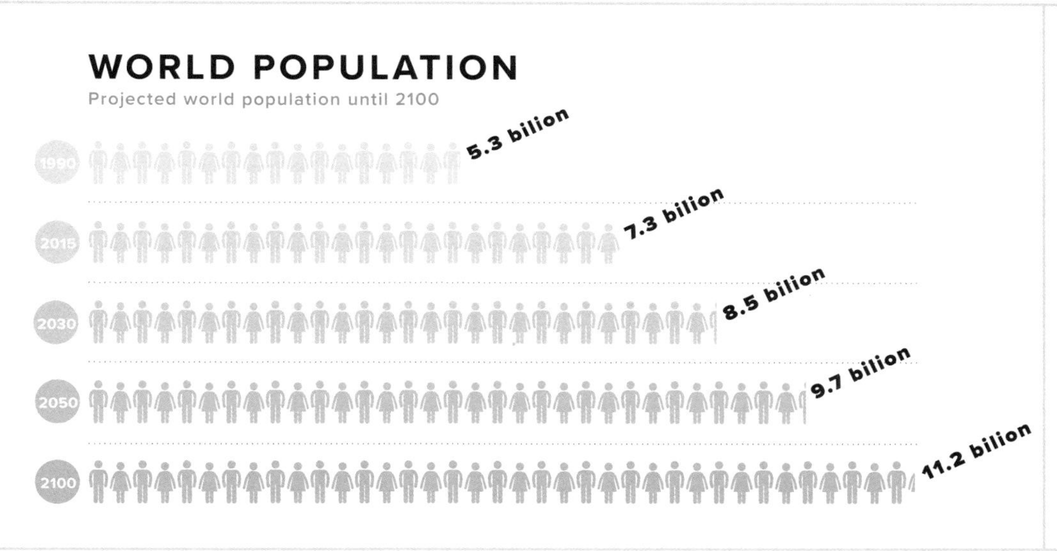

FIGURE
World population
Source: United Nations Department of Economic and Social Affairs.

In cities, the same level of exponential growth has taken place in just a single generation. Today, there are now more than thirty mega-cities with more than 10 million inhabitants. New York and Shanghai spring instantly to mind but cities like Lagos in Nigeria (20 million) and São Paulo in Brazil (22 million) are now also on the list. It is expected that by 2050 some 70% of the world's population will be living in cities. Providing food to these tightly packed urban concentrations of humanity will be more than just a major logistical challenge, not least because production rates – and therefore consumption rates – are higher in the cities than in rural areas.

For this reason, population pressure forms a challenge to our planet that is every bit as fundamental as climate change. In particular, it poses a massive problem for food security. With population growth on the scale anticipated, consumption will also increase exponentially, with all that this implies for waste, greenhouse emissions, the exhaustion of natural resources, etc. To feed the world's soaring population, the environmental impact of food production will increase by an estimated 50 to 90% between 2010 and 2050, even though the planet has already reached the limits of what it can reasonably sustain.

High concentrations of population also put other resources under pressure. For example, there are frequent riots in São Paolo because of water shortages. The deforestation of the Amazon region, primarily to provide fodder for livestock, has created a shortage of water throughout South America. As a result, not everyone now has access to clean water. But this is not just a problem for developing countries. Some predictions suggest that even a city like London faces the risk of water shortages in as little as 25 years. On the positive side, cities do still have the possibility to make their consumption more efficient, by shortening the food chain and by better coordinating supply and demand.

More, but also greyer

The massive growth of the world's population is not being caused by more babies. In the West, fewer babies are now being born than in the past. No, the main cause is a significant reduction in death rates. In other words, people are living longer. At the present time, each woman on Earth gives birth to an average of 2.4 children. The figure required to maintain the population level is 2.1 children. In many developed countries, the birth rate is well under this figure. In Great Britain the average is 1.7; in Germany it is just under 1.6. Put simply, in these countries not enough babies are being born to replace those at the other end of the age spectrum who are dying.

The United Nations expects that this curve will start to flatten around 2060. In fact, we are already witnessing the early signs of the first ever slowing down of population growth in human history. Of course, global problems of one kind or another can always reverse this trend. The lessons of the past show that the birth rate usually increases after major disasters, whether they are natural (pandemics, hurricanes, earthquakes, etc.) or man-made (famine, war, etc.). This helps to explain why population growth is currently strongest in African countries, which are hardest hit by such disasters. With just 7% of elderly people, Africa has a very young population, which therefore produces plenty of new children. Yet even in Africa the death rate is falling, so that the elderly group is now expanding. This growth in numbers at both ends of the life cycle means that countries like Congo and Tanzania are expected to see a tripling of their population by 2100.

Even so, the net result of all the above trends is that the population is getting greyer and greyer. Today, more than half the people in the world live to be older than 80 years of age. By 2040, a huge increase in the number of 'senior citizens' is anticipated. This is the generation of the baby-boomers. Born in the post-war period between 1946 and 1964, they will gradually fade away during the coming decades, a whole top layer of the population pyramid that will disappear. But until that happens, they constitute a high-care, high-cost group of some size and significance.

There is still much discussion about how old human beings might potentially become. The current record is 122 years. Be that as it may, it is already clear that in general we are getting older more healthily. New medical breakthroughs in the field of genetics have even led some groups of doctors to suggest that it may be possible to halt the ageing process. Until now, the gradual decline that eventually leads to death has been one of the few certainties in life. But perhaps no longer. This theoretically makes it possible for people to choose how long they want to live in good health.

Consumption behaviour changes as a person gets older. In Japan, which is the greyest country in the world, research has shown that the food expenditure of the elderly decreases dramatically. In particular, the consumption of fish has fallen significantly since 1990, notwithstanding a general rise in the country's prosperity. However, this does little to alter the fact that many other studies suggest that food and diet can be valuable allies in health care and play an active role in allowing us to enjoy a healthy old age.

Developing countries want their slice of cake (and meat)

The growth in population is not the only factor putting the food system under increasing pressure. Levels of development in countries throughout the world also have an impact. The idea that all will be well if we consume just a little less and are a bit more careful about what we eat is the classic example of white privilege. 'It is not only immoral but also untenable to think that the low-income countries, particularly those in Africa, will continue at their present levels of consumption. If you see how far these levels need to rise simply to achieve an acceptable minimum standard, it is clear that this will only be possible if our level of consumption in high-income countries is reduced,' says Sarah Harper, professor of gerontology at the Oxford Institute of Population Ageing (Packham, 2020).

When their income starts to increase, people in developing countries are likely, for example, to want to eat more meat. In Sub-Saharan Africa people currently consume a fraction of the recommended quantities of dairy products, vegetables and even chicken, at least according to the planetary health diet of the EAT-Lancet think tank. Using this same EAT logic, people in Southern Asia will need to eat one-third more red meat and cereals, while North Americans consume six times too much red meat and need to cut their intake of poultry by half.

This trend is already clear to see. It is the developing countries that are now pushing the worldwide consumption of meat, which increased by 58% between 1998 and 2018. These countries were responsible for 85% of this increase, with China leading the way. An estimated 72% of Chinese began to eat more meat during this period, accounting for one-third of total world growth. The main factor behind this overall growth was the rapid rise in the world's population during those two decades. Even so, almost half was also the result of increased consumption per capita. In view of the high ecological footprint of meat production – water consumption, greenhouse gas emissions, use of agricultural land – this correlation inevitably ramps up the pressure on our food system. On the plus side, there does at least seem to be a ceiling to meat consumption: in developed countries the growth in consumption is much slower and Japan has even witnessed a historic decline (-3%).

UNFAIRLY DISTRIBUTED

The United Nations has warned that we are still a long way from achieving a world with greater equality. Following a decade of improvement, the number of people suffering from hunger has risen again since 2015. One person in every nine is undernourished, while the UN's objective was to eradicate hunger by 2030. The corona pandemic threatens to double the amount of world hunger, as a result of a persistent economic malaise, inadequate access to food, and volatile prices and supply.

Poorer countries and low-income classes have been hit disproportionately hard by the corona crisis. Not only economically, but also in terms of health risks. Running parallel with rising hunger, obesity is also on the increase in every part of the world – and obese people are a favourite target of the coronavirus. Contrary to popular perception, obesity is not a problem associated exclusively with prosperity. For example, children in some of the most deprived regions in England are three times more likely to become obese (Dimbledy, 2020). According to the UN Food and Agriculture Organisation, people with little money generally eat poor quality and fattening food. The causes of obesity are primarily the same as for malnutrition: war and poverty (Dongyu e.a., 2020).

Food producer Danone says that it is already seeing food polarisation in practice. As a result, the former chairman of the company's board, Emmanuel Faber, predicts that the middle class will soon disappear, with the corona crisis as the catalyst (RetailDetail, Aug 2020). The gulf between rich and poor is growing, which means that so too is the gulf between the way people eat. In the new normal, some prosperous people and regions will spend more on healthy food, while less fortunate people and regions will be forced to save money on their diet. According to Faber, this polarisation is likely to get worse, resulting in economic and social crisis on the one hand and a growing interest in healthy food on the other.

In consequence, the risk of political unrest and destabilisation also increases. Scarcity and inequality always lead to poverty and polarisation. One reflection of this is the political discourse based on scarcity and fear that has gained ground in Western countries in recent years. Extreme right-wing positions, protectionist reflexes and intolerance are all growing in parallel with the increased pressure on natural resources. To complicate the situation still further, the United Nations predicts that in the immediate aftermath of the corona crisis a new wave of mass migration will occur, if hunger and economic crisis remains unchecked.

COVID as the catalyst

'Making peace with nature is the defining task of the 21st century,' concluded UN secretary-general António Guterres at the end of the crisis year 2020. The corona pandemic has succeeded in dislocating the entire world for the very first time. The virus exposed the vulnerability of our systems, including the food system.

The crisis was also a powerful wake-up call that accelerated a number of tendencies that had already been evident for some time. Not least in the food domain, corona was a 'perfect storm' that forced us to look at every link in the food chain.

If the system suddenly collapses

Today's globalised and industrialised food system is extremely vulnerable. That was the conclusion of a scientific study commissioned by the United Nations and the World Bank in the wake of the 2008 SARS epidemic. Further research conducted by the World Health Organization in 2015 broadly repeated the same message: 'Changes in land use and food production practices are amongst the most important catalysts for the development of illness and disease in people' (Herren, 2020).

In particular, corona again emphasised the fragility of the food distribution system. The food industry was pushed to its limits during the crisis in its attempt to keep the world's consumers provided with life's essentials in the right place at the right time. During the first wave of infection in the spring of 2020, this resulted in the fall-off of global transport, the closing of national borders and the slowing down of production due to the imposition of anti-corona measures. This was a massive stress test for the supply chains, the retailers and (not least) those working in the food sector.

In the West, stockpiling consumers were repeatedly faced with empty shelves, while elsewhere huge supplies were held up and unable to reach shops and supermarkets. Local purchasing soon became a necessity, highlighting our over-dependence on global supply networks. E-commerce in food was no longer a luxury for the few, but became a lifeline for the many. During the lockdown, trying to find a time slot in the click-and-collect service of your local supermarket became the equivalent of trying to find tickets for a sell-out concert.

The consumer explores new pathways

For consumers, the pandemic returned the kitchen to a central position in their lives. This increased attention to food was more than just a temporary lockdown effect. The role played by food in the interrelationship between sustainability and health is likely to lead to a permanent shift towards more aware consumption – at least for those who can afford it. In the fight against obesity and illness, governments, retailers and producers all now increasingly offer incentives to eat more wisely. In reality, they have little choice, since their ambitious climate objectives make it essential for them to 'green up' their range of products and to invest in sustainability.

In the urbanised world, the corona crisis showed how things can be done differently: the most significant retail change generated by COVID-19 is probably the breakthrough of e-commerce in food. This digitalisation of the food chain will continue to influence both supply and demand. Food is becoming a service, and preferably one that is close to where you live and work. Subscription services are flourishing – during the Christmas period, HelloFresh almost went into meltdown (RetailDetail, Dec. 2020) – and chains are being shortened.

In China, a huge battle broke out between the e-commerce giants to service the grocery needs of consumers at home. Some of their price actions were so aggressive that at one point the Chinese government became concerned. The companies squandered millions in their attempt to become the biggest player in a loss-making market, with the aim of forcing classic markets and food stores out of business (Tobin & Matsakis, 2021).

URBANISATION UNDER THREAT?

In London, the corona crisis resulted in a 71% decrease in the use of public transport. In Sydney, the figure was roughly one-third and in New Delhi it was 43% (Chandran, 2021). If more and more people work from home and therefore commute to the cities less and less, it seems likely that public transport will lose an important part of its function. Without a daily inflow and outflow of office workers, many cities risk becoming 'doughnut-shaped' (Hubbard, 2020): the centre – where the business districts and shopping streets are concentrated, but where no-one lives – is in danger of being hollowed out.

It is possible that some cities will never fully recover. The demand for properties in the countryside and homes with a garden has increased since the international lockdowns were enforced. If we are compelled to stay 'at home', we wish to do so in green surroundings where it is easier to maintain social distancing. In this respect, COVID-19 was very much an urban crisis: 90% of infections were recorded in cities (Guterres, 2020).

As in so many other fields, the pandemic has forced cities to look critically at themselves to see how they can become healthier and more sustainable systems. Following the example of Paris and Barcelona, the idea of the 15-minute city is gaining in popularity. Anne Hidalgo, the mayor of Paris, wants to make it possible for the city's inhabitants to fulfil all their basic daily needs within 15 minutes' travel on foot, by bike or with public transport. The aim is to reduce pollution, optimise mobility and improve the quality of life.

This means that the local district will be rediscovered and reshaped, with greater attention not only to local infrastructure, green zones, cycle and footpaths, etc., but also to local retail, markets, trading points and service outlets. In 2020, it was the local stores that enjoyed the biggest boost in growth. Similarly, smaller commercial centres recovered more quickly than their larger counterparts. In this way, a virtue was made of necessity: the importance of everything local grew significantly, whether from sustainability considerations or from more practical ones.

Likewise, local delivery services were increasingly seen as a solution, often specifically linked to a particular district or neighbourhood. In China, group purchasing has reached new heights since the crisis, especially in cities or parts of cities that are less easily accessible. Local residents frequently provide 'last mile' delivery in return for a small commission. In the meantime, the Dutch online supermarket Picnic is extending its activities to France and Germany.

How Paris wants to become a 15-minute city is illustrated in this short video:

https://youtu.be/MUj_QDfEgPc

The consumer is hungry for tomorrow

Mega-trends in the 'transition twenties'

Society is constantly in movement. Not only have we started a new decade, but according to some trendwatchers the 2020s will also signal the start of a new era. During this period of ten years – the so-called 'transition twenties' – a number of far-reaching changes will take place. COVID-19 was just the beginning ...

The most important driving forces behind these revolutionary developments? Climate change and artificial intelligence, which together will cause a systemic shift (Toch 2020). Entire systems will need to be reconfigured, not least the food system.

Progress, however, never happens in a straight line. One of the characteristics of a world in motion is friction, the field of tension that occurs between innovation and conservatism, between protecting what was and being open to what is yet to come. Every trend generates a countertrend and, like every other great change in history, this new evolution will be marked by action and reaction.

You get a clear sense that in everything, from politics to purchasing behaviour, the transition consumer is constantly searching. The fact that during the corona lockdown many consumers turned to the snacks and sweets of their youth as comfort food does not mean that plant-based and healthy eating are not growth trends. Notwithstanding a number of paradoxical and even downright contradictory phenomena, a long-term evolution is unquestionably taking place. This explains, for example, why 30% of Europeans say that they want to devote more attention to healthier food, whilst at the same time the same number – roughly one in three – say that they plan to save money on their groceries budget.

A fluid society

Modern-day society is fluid. Things that are 'standard' are no longer in fashion. The days of 'one size fits all' are over. Each day, our pigeon-hole thinking is slowly being eroded. Even though this evolution does not always run smoothly or at the same speed, Western consumers are increasingly profiling themselves as 'unique' individuals with highly personal characteristics, wishes and needs.

In this process, the power of the internet as a connector is crucial. People not only have the possibility to gain information and form an opinion about absolutely everything, but will nearly always find others who share that opinion and will almost always be given a platform somewhere – even if it is only in the comments section of an online newspaper. No matter what you think or do, in today's world you are never alone.

Although, as a result, it sometimes seems that the world is more divided than ever, it needs to be remembered that trolls and hate-mongers are just one side of the coin. Worldwide connectivity also offers huge potential for uniting and emancipating people, offering them the freedom to discover and be themselves.

The youngest generation, Gen Z, are the clearest example of this. The common denominator of people born between 1996 and 2012, according to a research study by McKinsey, is their desire for truth and veracity. Working independently, various research teams in Asia, the US and Brazil have concluded that this generation expects personalisation, is searching for uniqueness, and gives a high priority to personal expression. This is a crucial fact: the generation to which they belong says more about their behaviour than socio-economic or geographical differences, even across different continents. They are growing up together, in a shared and connected world.

For Gen Z consumption is a medium to express their individual identity – and that identity is fluid. These 'identity nomads' avoid labels and refuse to let themselves be pinned down by stereotypes. Perhaps it is a characteristic of their youth, but they see it as their task to experiment with different ways of being who they are. As a result, they are less inclined to pass judgement on others: everyone has the right to his or her own truth. In their search for authenticity, they are strongly in favour of freedom of expression and greater openness to the understanding of different kinds of people.

Gender fluidity is probably the most striking example of this, but by no means the only one. The importance of age has been superseded by the phase of life in which someone finds themselves, the income category to which they belong, and the cultural background from which they come. Because people are not only living longer but also getting old more healthily, the concept of age is being redefined in a more individual way. This is reflected in the need in some countries to find a new word to describe a new and increasing phenomenon: the active older person, who might well be 'senior' or a 'pensioner', but is still in the flower of his or her life. Hence the emergence of terms like 'perennials' (USA) and 'jagger' (Belgium).

MAN IN A SKIRT

Robot engineer Mark Bryan became famous on social media as the man who wears skirts. Bryan identifies himself as a heterosexual man who loves typically 'masculine things' like beautiful women, the football team he coaches, his grandchildren and fast cars. But when he goes into the office, under his (male) shirt and tie he always wears a pencil skirt and high heels. He believes that clothes should not have a gender.

Uniformity is unappetising

What is the impact of these developments on what we eat and how we do our grocery shopping? For decades, food producers and supermarkets have focused on reaching the greatest number of people with the fewest number of products.

This is known as category optimisation. The supermarket – or, preferably, hypermarket – was supposed to be the ultimate one-stop shop, where everyone could meet as many of their grocery needs as possible. The result was an offer that was uniform and uninspiring.

Today, a new model is urgently required. Classic segmentation no longer works, because it gives a distorted and one-sided image of the true situation. Fortunately, producers and retailers now have much more and much more relevant consumer information at their disposal. This data makes personalisation possible, but without losing sight of the importance of the economies of scale of global supply and distribution chains.

Because consumers nowadays are constantly in (digital) contact with the entire planet, the range of goods within their reach is now almost limitless. The world is quite literally at their fingertips. If producers and retailers fail to take sufficient account of specific new needs and, in particular, demographic shifts, they can be certain that specialised challengers and alternative disruptors will fill the resulting gaps in the market. In an interconnected world, no niche is too small.

HALAL FILTER

The Halal Food Shop is determined to become the largest online halal food store in the United Kingdom. Although premium producer Haloodies saw its online sales increase by 30% in 2020, primarily via Amazon Fresh and Ocado, the company still sees a need for a specific distribution channel that makes it simple and easy for Muslims to do their grocery shopping, a process that is still in many cases far too complex.

Of the classic British supermarket chains, only Morrisons has a halal filter across all its categories – a surprising omission when one considers that the British market for halal food is estimated at 4.5 billion pounds and that the Muslim population in the UK is good for 4% of the country's total expenditure on food and drink. It says much that it was only as recently as 2020 that M&S Food introduced its own range of ready-to-eat halal meals.

Building up the necessary trust is a slow process: in 2008, most British Muslims were still unwilling to risk buying halal meat from one of the large supermarkets. In fact, the majority did not even know that the supermarkets in their neighbourhood usually sold halal-certified meat products. Although the range of goods and people's awareness of them have increased in the interim, many Muslim consumers still prefer to trust the products sold by retailers who are themselves Muslim. As a result, mass distributors and brands are missing out on the purchasing power of 4.5% of the British population.

The immaterial society

Now that your evening meal can arrive on your table with a just few clicks of your smartphone and the menu for the entire week can be delivered to your doorstop each Monday with equal ease, it is truly possible to say that society has entered a new phase. A digitalised, service-oriented and even immaterial phase. Immaterial, because an increasing number of primary needs are met automatically, without the necessity for consumers to actually go and buy things physically.

By 2030, all aspects of basic domestic consumption will be arranged via subscription systems, predicts the Copenhagen Institute for Futures Studies. This will even apply, they claim, to things like clothes and cars. We already find it 'normal' that music and film are no longer provided to us in material form but through streaming. So why should we not extend this same ease of purchase and use to other needs? In Belgium alone, the demand for meal boxes increased threefold in 2020.

In particular, it is the younger generations who are increasingly attaching more importance to use and experience than to possession. The consumer society – in reality, an over-consumption society – has reached its saturation point in the

West, as a result of which the focus is shifting towards immaterial values and service provision. In recent years, this has been evident in the move towards an experience economy, even in the retail sector.

In this post-materialist context, the consumer is more inclined to search for services and solutions than for products. The busy Westerner now wants things that will make his life easier, save him time or offer him a memorable experience. The easier and more automatically these things can be arranged, the better he likes it. This phenomenon is further strengthened by the preference for a contact-free economy and social distancing that became commonplace during the corona pandemic.

Not physically, but online

It is immediately noticeable just how closely this trend is interwoven with digitalisation. New technology makes possible new ease and new experiences. The year 2020 saw the real breakthrough of e-commerce in the food sector: one out of every ten European consumers now does their weekly grocery shopping online. In digitally mature markets, like the UK and the Netherlands, this figure is already higher than 30%, although increases are also evident in less digitalised countries, such as Italy and Spain.

In ten years' time, it is probable that things will have progressed so far that all purchases will be made automatically and without thinking online, so that physical purchase then becomes a matter of conscious choice. In other words, digital consumption will define physical consumption, and not the other way around (Ulrik, 2020). Put simply, this means that the need for physical shops will decline, as will the willingness of consumers to accept inefficient or unpleasant shopping experiences. A lack of flexibility or an unnecessary loss of time will not be tolerated. The amount of time spent on digital devices has increased exponentially in recent years, thereby reducing the amount of time available for analogue experiences. By the same token, demanding consumers will also expect physical stores to provide the same level of ease and service as their online counterparts.

In future, every interaction will need to be worth the consumers' investment of their valuable time. This will apply equally to the food sector, from which much in terms of ease and experience will be expected. The vending machines of Let's

Pizza already bake fresh (not frozen) pizzas in front of the customer's eyes in just three minutes. The dough is mixed, kneaded and bakes inside the machine, without the need for any kind of human intervention. By 2030, the CIFS predicts serious competition for the traditional small pizzerias from fully automated machines of this kind, which not only prepare contact-free pizzas, but also deliver them to your door in self-driving vehicles, and all for a fraction of the price.

Meeting your meal

Of course, not everyone will want to order all their pizzas from robots. Real pizza lovers will still prefer authentic hand-kneaded, oven-baked pizzas in atmospheric Italian restaurants. It is all a question of what is really important to the individual consumer. Is it speed and ease? Or is it flavour and experience? For this reason, it will continue to be crucial to make a distinction between products with a high and a low degree of engagement. Products and services that are unable to engage or fascinate consumers will need to focus on further automation and digitalisation, because for them 'convenience' will have to be their trump card.

For products with a high level of engagement, consumers will still be willing to make a greater effort. Fortunately, food in general falls into this latter category, although retailers must not forget that this does not apply to every item in their range: people will never be excited by toilet paper, cleaning products or tinned peas! In practice, this means that it will be necessary to develop two streams within the same shopping format and even within the same shopping visit: a hyper-efficient 'easy-as-possible' stream and an experience-oriented service stream.

This increased focus on experience implies that people will also want to 'experience' their food (immaterially), instead of merely 'consuming' it (materially). They will want to know and understand their food, get a real sense of what it is. Naturally, this is closely related to cooking culture and the expectation that food must be something more than just a way to fill your stomach – it also has to be 'pretty' and 'good' – but it is just as much a part of the new experience economy. Supermarkets are converting their fresh-food sections into open marketplaces, where shoppers can follow workshops, learn more about the third-world farmers who grow their pineapples, exchange recipes and cooking tips with fellow shoppers, etc.

Traditional crafts are also making a comeback – even if only as a marketing term to which lip service needs to be paid. Or is something more going on? Anyone who has ever baked their own bread knows just how much work is involved and is therefore perhaps more likely to appreciate the skills of the baker better than ever before ...

The aware society

As in so many other fields, consumers are becoming more aware and, consequently, more critical of what they eat and drink. They are starting to better understand that their diet has an important effect on their health and well-being. This insight makes it possible for them to make more conscious dietary choices and results in more empowered consumers.

Even before COVID-19 overwhelmed the world and forced people back into their kitchens, a strong food culture already existed. However, the pandemic has made the link between food, health and sustainability that much clearer. For today's consumer, food has become something more than just food for the body. It is now food for the soul. Instead of products, men and women everywhere are searching for the solutions and services that best match their way of living and support them in their quest for a better life.

They expect brands to become allies and for retailers to become advisers. Smart retailers and brands respond to this expectation. During the pandemic, for example, the American supermarket chain Kroger launched a free tele-nutrition service. Dietary specialists gave free virtual coaching sessions to all Kroger customers. The Kroger Health Team consists of no fewer than 22,000 care professionals, ranging from dietists and chemists to nursing personnel.

Food is intimate

Food is functional, but has also become something intimate and very personal. Consumers now expect 'customised' food solutions that satisfy their unique needs and preferences. In this way, food has also become a vehicle for health and well-being, two aspects of life that are increasingly driven by data and technology. For example, the app and the smartwatch of DNANudge tell you what you should buy and what you should avoid in the supermarket, based on your personal DNA. The app was developed with the support of the British Department of Health.

In recent years, Nestlé has taken over both LivingMatrix and Persona: two companies that are specialised in functional and personalised foodstuffs. LivingMatrix makes use of data technology and algorithms to support doctors in the development of individual care plans for people with chronic illnesses. Persona makes personalised vitamin and nutritional supplements. In Japan, Nestlé is already offering consumers personalised food products based on their DNA.

In view of the great importance that is attached to food in this new landscape, consumers now make exacting demands in terms of quality and transparency. Because these are sensitive issues, winning consumer trust is crucial for the food industry. According to the European Union's Fit4Food 2030 project, the public evaluates and assess food products and food technologies in a number of different (and sometimes unexpected) ways. These associations are often emotional, social and based on prejudices.

With each new food scandal – and in our information age more and more of these are coming to light – public confidence in the food industry takes another battering. There is a growing realisation among consumers that food is not safe per se, never mind responsible. This is reflected in their choices and purchasing behaviour: today's food must be good, clean and fair.

Shopping on the basis of your DNA? The founders of DNANudge explain how it works:

https://youtu.be/E8CIx0ONLGs

Food must be good

First and foremost, food must be good. But the meaning of 'good' in a food context is evolving at lightning speed into something like 'good for me and my body'. In other words, good means healthy, high-quality and delicious. Since the Western world discovered kale smoothies and quinoa salads, the health trend has become a part of our modern way of life – and a part that is here to stay.

According to most players in the food industry, plant-based meat and dairy substitutes are the next big market opportunity in the sector, thanks to the combination of health and ecological benefits they provide. For example, Danone wants to more than double its sale of plant-based products from 2 billion to 5 billion euros by 2025. Within this growing market, there is one item that is universally regarded as the holy grail: the race to develop the tastiest, juiciest and meatiest vegetarian burger. This race is already in full swing and a recent court case costing millions underlined just how much is at stake: Nestlé was forced to drop the name of its brand-new 'Incredible' Burger for no better reason than it was judged to sound too much like the 'Impossible' Burger of one of its main rivals.

The health trend was another phenomenon that was given a serious boost by the corona crisis, since it reflected people's growing desire to live a long and healthy life, free from medical complaints. This, in the opinion of food scientist Hanni Rützler (2020), is a natural evolution during a period when things are going wrong and everyone wants to rediscover their former sense of stability and security.

The soft or hard approach

That being said, Rützler makes a distinction between what she calls 'forced health' and 'soft health'. In the first case, consumers eat particular foodstuffs because of their perceived preventative or curative health effects. Whether or not they are tasty is of secondary importance. For example, since the start of the corona crisis there has been a renewed demand for nutritional supplements, even though this is a phenomenon that originally dates from the 1990s.

In this respect, there is now much more attention to and much more knowledge about allergies and the functioning of the intestines. Scientists at the Catholic University of Leuven have recently discovered that dietary changes are more effective for treating irritable bowel syndrome than medication. Consequently, they have developed an app that will help individual patients to make the right dietary choices. Moreover, there is increasing research evidence that intestinal problems can cause depression, anxiety conditions and even autism (Svoboda, 2020).

At the commercial level, the corona pandemic again (and not surprisingly) gave a significant boost to the market for 'immunity-strengthening' products. In the middle of the pandemic, Upfield launched a new variant of their Becel margarine, which was supposed to strengthen the immune system through the added vitamins it contains. Becel had previously targeted fifty-plussers suffering from high cholesterol with its ProActive cholesterol-reducing variant. The new product is designed to appeal to a younger target group.

On the other side of the dietary equation, a softer and more holistic approach to food and drink is also gaining ground, in which the number of calories or the precise nature of the constituent elements are no longer the determining factors for healthiness. In this trend, consumers above all associate fresh and plant-based foodstuffs with good health, while in general they also regard locally sourced food as being better and healthier.

The growing demand for organic and local food is directly related to the recent spate of food scandals and the negative publicity that industrialised mass production has increasingly received. Listeria and E.coli outbreaks, the appalling conditions in many slaughterhouses, fraud with ingredients (do you remember the 2013 horse meat scandal?): these and other similar incidents have all helped to ensure the rapid spread of organic and local products (Wepner, 2021).

Public opinion generally seems to believe that there are fewer health risks associated with organically cultivated produce (although an increase in the number of fraud cases is starting to eat away at the public's confidence). In particular, consumers assume that organic food offers a solution for the diseases and deaths caused by the use of pesticides. As concerns about food quality and sustainability grew with the corona pandemic, so too did the demand for shorter supply chains and, consequently, local products.

In 2020, roughly 30% of Europeans said that they intended in future to devote more attention to a healthy diet. More than a quarter of them are willing to spend more on regional and local products. A further 19% are planning to buy more environmentally-friendly products (Gerckens, 2021). In Belgium, these trends are even stronger: 36% of the population associate plant-based foodstuffs with health.

Food must be clean

Just 5% of the European population need to avoid certain foods for medical reasons. Even so, allergy-free food is one of the fastest growing categories. Increasing concern about food allergies and intolerances not only reflects their perceived nature as so-called 'civilisation' or 'prosperity' diseases, but is also an expression of people's new way of dealing with food and the search for a new and healthier dietary balance that this involves.

Modern consumers are less and less inclined to trust the industrialised food system, leading to a flood of criticism about over-processed products with too many (artificial) ingredients and substances. In contrast, natural foodstuffs – from keto through vegan to raw – are now generally regarded as being 'clean'. The 'paleo-hype', in which people only ate the things that our cave-dwelling ancestors would have eaten, is perhaps the most striking example of this phenomenon.

When searching for security in our high-risk society and for purity and hygiene in a post-pandemic era, the majority of consumers regard products that can trigger allergies and intolerances in others with deep suspicion. This same suspicion also extends in many cases to preservatives, additives and flavourings. These are now generally seen as 'things to avoid', whereas the terms gluten-free, lactose-free, fat-free, sugar-free, etc. are associated with health, weight loss and a natural way of eating.

The importance of safe, healthy, nourishing and authentic food has become a permanent theme in our eating culture. As a result, transparent and open information is now a necessary prerequisite for all producers. The more complex and more global the food system becomes, the less faith consumers have in its trustworthiness and the greater the need for reliable information about the origin, content and processing of the food that is produced. Today, this kind of information is still often a unique bonus or added extra for consumers, but in the future data technologies such as the blockchain and RFID tag tracings will make this information the new norm.

In this context, the importance of food and storytelling is also growing. The 'Noma effect' (Rützler, 2020), named after the famous multi-star restaurant in Copenhagen, has helped to focus attention on (hyper-)local food, biodiversity and seasonal variations, not so much from the perspective of rational argument and awareness raising, but as part of a process of sensory and emotional experience. As a result, a number of forgotten vegetables have been resurrected from obscurity and elevated to the status of super-chic haute-cuisine by famous chefs who are increasingly keen to portray themselves as influencers rather than just cooks.

WHAT YOU DO FOR YOURSELF YOU ALWAYS DO BETTER

What can give you more trust and certainty than something you make yourself? The phenomenon of the 'hobby cook' has been with us for years, but it was yet another existing trend that was given a significant boost by the corona crisis. More than 70% of European consumers want to continue the new cooking passion they discovered during the pandemic, as a result of which they plan to eat out less and cook more at home. This means that in the 'new normal', a balance will be found between food retail, the catering industry and other options within the food ecosystem.

This cooking culture is an essential part of the importance that today's consumers attach to food and drink, not only for culinary reasons, but also in relation to lifestyle and experience. By working in their garden and cooking for themselves, many Western consumers hope to slow down the hectic pace of their life and find a new way of living that is purer and brings them closer to nature. In this way, making your own jam or bread again serves as a counterbalance to the complex, technological world. Viewed in these terms, cooking becomes an activity in its own right rather than being no more than a means to an end.

As a result, modern cooking culture is closely related to people's search for health, sustainability, authenticity and even a more ethical way of life. It also helps to satisfy emotional and social needs through its strong connection with social media, the online and offline sharing of cooking tips, and the increasing formation of communities based on food and dietary choices. If you are a vegan or are gluten-intolerant, worldwide connectivity means that nowadays you can enjoy the support of a large international community of like-minded people.

However, cooking for yourself also implies growing and producing for yourself. Whether or not you do this consciously as part of the do-it-yourself trend around 'clean' food makes no difference: it is something that needs to happen – and the sooner, the better. The Fit4Food 2030 project of the EU sees a direct connection between cooking for yourself and phenomena such as 'pick-your-own-fruit' farms and urban agriculture. 'Whether it is growing flowers or harvesting vegetables, everything is possible. Do-it-yourself food is becoming a lifestyle, pushed via social media channels and food blogs, but also via the availability of simple instructions for pickling and conserving that can be found on the internet. Green zones in urban areas must bring people together, provide them with food and make their cities more attractive and more sustainable.'

This might sound small scale and somewhat nostalgic, but in the long term it can have a major positive impact. The wider direct-to-consumer movement, which cuts out the intermediary and sometimes even the professional producer, is well on its way to penetrating the food market. Bearing in mind the growth in short-chain initiatives, both online and offline (for example, farm shops), it is only to be expected that sales from consumer to consumer (C2C) will form serious competition for retailers and producers in the years ahead.

At the present time, the lack of a clear and effective legal framework still represents a significant obstacle to the breakthrough of C2C or the do-it-yourself market in food. However, there is growing pressure to implement the necessary reforms that will allow this flourishing trend to be removed from the grey area of the informal economy. What is the problem? As soon as someone offers food for sale that has not been grown for personal consumption, that person is regarded under European law as a food business operator and is therefore automatically subject to the same national and European legislation as the food professionals. This needs to change.

Food must be fair

The term 'sustainability' covers a multitude of meanings, ranging from environmentally friendly to healthy. For this reason, the term 'fair' allows a more precise identification of the key elements. 'Fair' food emphasises that the processes of production and distribution must not only be beneficial for the consumer, but also for the planet and people in general. In other words, fair food is an altruistic reflex.

Almost 60% of Europeans – irrespective of age, gender or other demographic characteristics – say that they would be willing to pay something more for sustainable food, even if 34% of them hope in future to save money on their grocery budget as a whole. The overall intention to buy sustainable and environmentally-friendly products has risen in recent times by more than 8% (Gerckens, 2021).

It is certainly the case that Gen Z's search for truth and veracity goes hand in hand with a degree of activism. Seven out of every ten young people think that it is important to defend everything that is connected with individual identity; as a result, they are also concerned about human rights, working conditions and ethical questions.

Gen Z (60%), shoppers with a higher income (60%) and women are the people who are most willing to pay more for fair and sustainable products. Even though there is inevitably a difference between what some people say and do, these figures nevertheless suggest that in practice sustainability is economically viable: the sale of consumption goods that are marketed as sustainable is growing almost four times faster than the average. In other words, there is a correlation between a company's sustainability policy and its financial results.

Even so, it needs to be borne in mind that not every sustainability objective is regarded as being important by everyone. Consumers consider sustainability to be important for fresh produce (fruit, vegetables, fish and meat), but what most motivates them to pay more is the promise of better working conditions and pay for the people employed throughout the value chain, which was the original meaning to the term 'fair trade'. There are also a number of environmental considerations for which consumers would be prepared to pay more, such as products with a low rate of greenhouse gas emission or those that are free from harmful substances, such as palm oil or micro-plastics.

THE COUNTERTREND: NOT EVERYONE IS 'FOOD LITERATE'

Not everyone knows the difference between healthy and unhealthy food, or else cannot afford to make the distinction. The level of inequality between different groups of people is large and the development of current trends seems likely to make it even larger: on the one side, there is a growing group of people who demand higher food standards and are willing to pay for them; on the other side, there is an even larger group who are struggling financially to make ends meet and are neither willing nor able to pay more.

Although the eating patterns of people with a low income and people with a high income are broadly comparable, the diet of low-income families displays a number of common 'unhealthy' characteristics. People with a low income tend to eat less wholemeal bread and fruit, while they eat more fats and oil, sugar-based soft drinks, pizza, processed meat and refined sugar.

People who have less money, are less well educated and have poor living and working conditions often opt from necessity for cheap foodstuffs with a high sugar and fat content. Nowadays, there are plenty of these cheaper food options available, but they often contain more 'empty calories', which eventually results in an increase in obesity within these socio-demographic segments. In developed countries, obesity – one of the main causes of non-transferable diseases – is over-represented in the countryside and among poorer communities. In this way, the vicious circle of poverty, unhealthy diet, obesity and disease is perpetuated. The European Fit4Food project therefore recommends that further efforts are needed to promote food awareness and maintain a correct price balance, so that healthy food remains accessible for everyone.

The trust economy

Trust is everything. Trust determines whether or not consumers remain faithful to a brand and also the extent to which they are prepared to take risks or make an effort on its behalf. The highly personal and primary role played by food means that trust in the food industry and food retail is doubly important. Are consumers prepared to go the extra mile for your product, to pay more for it and to keep on buying it? Congratulations: they trust you!

Trust has also been shown to be an important factor in persuading people to change their behaviour (Rampl, 2012). You want people to eat more healthily? This will only be possible if they believe that the food you are selling them is indeed healthier: that the health claims on the packaging and the list of ingredients are 100% accurate and honest. The same applies to sustainability and quality.

Unfortunately, consumer trust is waning. In 2008, the Tesco supermarket chain was the organisation most widely trusted by the British public, even more so than the British government. During the financial crisis of that year, people felt that the supermarket was more ready to help them than their elected political leaders. In contrast, ten years later the food industry, together with the car industry, were the two sectors that had most lost the public's confidence. Since then, declining consumer trust in the food chain has been a constant theme.

Because of the ever increasing distance between food production and consumption, both geographically and in terms of time, consumers are feeling more and more uncertain about the authenticity of what they eat. Food scandals and crises, which are made public more often and more quickly in our age of information, serve to strengthen this feeling of alienation and distrust. Consumers today understand less about where their food comes from and how it is produced – and that is not a good thing for promoting trust.

Transparency

International scientific research has concluded that there are three key factors for generating consumer trust: competence, carefulness and openness, with openness being the most important of the three (Macready, 2020). Above all, it is the openness of producers that has the biggest impact. In comparison, the competence of retailers weighs less heavily in consumer thinking. As far as sustainability is concerned, the public only attaches importance to the role of the farmers and the manufacturers.

In other words, transparency is key. One of the side-effects of worldwide connectivity in our internet age is that consumers expect instant access at all times to correct, complete and relevant information. Not only about quality and product characteristics, but also about price and data use. Consumers have become so used to being able to compare that they now expect something in return for the use of their data. Openness works both ways.

Modern consumers know exactly what possibilities their data offers to companies. For example, it allows the website to be adjusted to the preferences of the individual visitor; it makes possible personalised advertising; it facilitates the build-up of a customer's purchasing history. Most people do not have a problem with this, but expect to be 'rewarded' in return. Companies need to continually prove that they are worthy of their consumers' trust and that these consumers also benefit from their openness. No fewer than 60% of American consumers distrust the accuracy of food labelling. At the same time, almost everyone (81%) thinks that it is important for food manufacturers to be transparent about the contents of their products and how they are made (FMI, 2021). For 70% of Americans the completeness of the list of ingredients plays a role in their purchasing decision. More than a quarter (27%) is prepared to change brand if an alternative brand offers more or clearer product information.

In short, transparency is a crucial determining factor for purchase behaviour and is getting more important all the time. Not simply from sustainability considerations (a concern for roughly one in four Americans), but also from the perspective of changed eating habits. As many as 64% of American men and women follow a fixed diet or make food choices based on health considerations, while more than half devote attention to product information about allergies and intolerances. What's more, these figures increase by around 10% each year.

What can producers, manufacturers and retailers do to increase trust? As early as 2016, consumers indicated that their highest priority was digital labelling, providing detailed information about allergies and nutritional value. In the meantime, more than half now go in search of more information online. Where possible, the vast majority of shoppers are also prepared to go online while they are in stores and supermarkets or else to make use of apps that give more extensive product information.

There is still room for improvement in these online services. Many consumers would like to be able to filter on the basis of specific product characteristics or dietary preferences – from non-genetically modified to keto – and would be happy to receive product recommendations on the basis of nutritional and ingredient information. Customers are willing to provide companies with their personal data, but in return expect better advice and more customised service provision.

The new network organisation

In our hyper-informed, hyper-connected world, trust often goes hand in hand with familiarity, with something or someone to hold on to amid the vast mass of available information. Who is there to help you, when you can no longer see the wood for the trees? Where is there an objective and reliable place that will summarise all this information for you cogently and clearly? Questions of this kind offer important opportunities for retailers, since they are the pre-eminent point of contact for consumers.

Modern consumers go in search of partners that they can add to their network and who they can trust to relieve them of some of their purchasing problems. They must be partners who share their values, who show themselves to be sincere and authentic, and who involve the consumer actively in their processes. The one-way traffic in the customer journey of years gone by has now been replaced by a circular trajectory, in which the customer is central and there are numerous micro-moments for contact and mutual influencing. In particular, it is important for the young people of Gen Z that brands share their values and set the same priorities. They expect to be listened to, in a dialogue that today takes place largely via social media.

As part of these personal networks, the distributor also simultaneously becomes a service provider. In their role as large retail stores, filled to the rafters with bulk goods, the supermarkets are faced with stiff competition from high-performance logistical e-fulfilment centres operating on online platforms, while consumers also go independently in search of solutions and service models that can act as allies in their busy lives.

The supermarket of the future must become a giant hub, where consumers can go to make use of a wide range of services, from food and drink to health care. In Normandy, the French Intermarché supermarket group has launched a first concept store designed especially with seniors in mind. 'Bien Chez Moi' is not a classic homecare store, but it does stock an extensive assortment of products for an older public, including books about grand-parenting and aids for home adjustments for those with limited mobility. Having said that, nothing in recent times better demonstrates the role of the food retailer in modern society than the use of Walmart hypermarkets as vaccination centres for the inoculation of millions of Americans against the corona virus.

This is a good example of how food outlets can (and must) take on a social role in a digitalised society. This, too, is a consequence of transparency. In a world where price-comparison apps and smart shelf labels make possible the instant adjustment of prices, so that price differentiation is becoming almost impossible to achieve, retailers need to find some other way to stand out from the crowd. Assuming, that is, that we will still have supermarkets of bricks and mortar in the years ahead: the growth of online grocery shopping seen during the corona crisis looks set to continue and expand even further, calling into question the possible future need for physical retail stores (Gerckens, 2021).

Before they can become part of a consumer network, retailers first need to turn themselves into a network of their own. It will be difficult to provide the wide range of services they need to offer without help. This can only be done through collaboration, by creating an ecosystem with the right partners who also enjoy consumer trust. Although this is very different from the way companies have operated in recent decades, in the future it will be necessary for them to look more outwards instead of inwards and to surrender a degree of their traditional control.

In the current competitive landscape, with unbeatable players like Amazon, the vast majority of retailers are likely to get much further through smart 'coopetition' in partnership with (former) competitors. Companies that want to stay in the game, both with the consumers and with their new challengers, will discover that openness is also crucial between partner organisations. Coopetition means that you stand apart at the front end, but collaborate closely at the back end. More and more retailers are already doing this, such as Fnac and Carrefour in France or Delhaize and Albert Heijn in Belgium. Pure competition, as we have known it for the past half century, no longer exists.

Take a look at Intermarché's 'Bien Chez Moi' seniors store:

https://youtu.be/u-KREU-jFY8

The relational employee

Robots and artificial intelligence have only one purpose in 'life': to become as human as possible. But nothing is quite so human as a human being. For this reason, the role of the shop employee in years to come will develop into that of a host, who builds up a close relationship with the customer. In this respect, the retail sector can learn many lessons from the restaurants and hotels of the hospitality sector. As a result, it will become possible to optimally combine the rational power of the computer brain with the emotional and creative power of the human brain.

While technology can work wonders for efficiency, the physical store as an anchor point has the additional potential to become a familiar place where customers feel comfortable and 'at home'; a place where the experience and the story of the brand can be conveyed to the customer through the different senses by shop assistants who are brand ambassadors par excellence. Their task is no longer pure selling, but consists instead of storytelling, service provision and the building-up of a relationship of trust with the customer. Because the technology of the future will soon be able to take over (or at least lighten) the burden of many practical tasks, more time will be freed up on the shop floor for the staff to perform their unique human role.

Repetitive work, which is of little direct value for the relationship with the customer, should, wherever possible, be automated. Since the outbreak of the corona crisis, food retailers have invested heavily in online fulfilment, which currently involves store personnel preparing online orders for collection or delivery, although even this is a task that can almost certainly be automated in due course. In some places, it is already happening. For example, the Amazon supermarkets make use of high-tech mini-warehouses where a large part of the order picking is done automatically. The Los Angeles branch of Amazon Fresh has a micro-distribution centre positioned around the edges of the shop space, equipped with smart conveyor belts that bring together all the non-cooled elements of the orders. The human personnel only need to add the fresh and frozen elements.

When people and machines share the same work space, there needs to be a degree of mutual understanding. Headquarters and operational staff will need to acquire new skills, like data analysis and coding. However, diversity and a multi-disciplinary approach will continue to be important. Even though accurate data predictions will be an indispensable operational aid, it is what the relational employee does with this data that will be most crucial to success. Can the assistant on the shop floor take advantage of his knowledge of the customer's preferences? Is he authorised and able to take specific action to respond to those preferences?

A strong technologically-driven backbone must provide precisely this kind of additional autonomy on the shop floor. Local teams must be empowered to take decisions on the basis of the data made available to them, adjusted to reflect their knowledge of the local situation. Once again, this is a question of trust: will local teams be given the authority to respond independently to the kind of personalised service provision that customers are increasingly coming to regard as standard? For the new wave of employees from Generations Y and Z, this is the kind of responsibility they want and expect to receive from their employer. To make this possible, open, transparent network structures with self-steering teams will systematically replace the hierarchical top-down approach of recent decades.

Ingredients for a new food system

The menu for 2050

Will our planet be capable of providing sufficient food to feed ten billion people healthily by 2050? Thirty-seven renowned scientists from around the world were asked to consider this question. Their answer was: 'Yes, but ...' In their opinion, it will only be possible if we drastically change our eating habits, if agriculture and food production is significantly improved, and if waste is limited.

Their more detailed conclusions can be read in the EAT-Lancet report, which was published in January 2019 with the aim of formulating scientifically grounded conclusions about what sustainable food production and healthy eating habits might involve. Admittedly, their findings were not without controversy, but they at least offered a useful framework for dialogue. The important thing is that science, the food industry and the retail sector are all working to try to find solutions.

What is a healthy eating pattern?

According to the report, a healthy eating pattern must optimise the health of the person concerned in the broadest possible sense. This means a condition of complete physical, mental and social well-being, and not just the absence of illness and disease. In concrete terms, the report argues for a plate of food that consists of roughly 50% of fruit and vegetables, with the other half being made up from whole grains, vegetable sources of protein, unsaturated vegetable oils and (optionally) a modest amount of animal protein, such as meat, fish or dairy products. The consumption of refined cereals, highly processed foodstuffs and added sugars should, ideally, be drastically reduced.

Compliance with these recommendations would require some serious changes in our eating habits. We would need to double our present consumption of healthy foods like fruit, vegetables, legumes and nuts, whilst at the same time halving our intake of less healthy foods like added sugars and red meat. These figures relate to a worldwide average, which means that the emphasis will have to be placed on reducing over-consumption in richer countries. In poorer countries, the populations do not currently enjoy the luxury of Western choice. In these latter countries, the demand for animal-based products will continue to rise, even if the West succeeds in cutting its own massive over-indulgence. Be that as it may, complying with the recommendations would result in an estimated reduction of some 11 million deaths each year, equivalent to between 19 and 24% of the total number of adult fatalities.

More sustainable production

If we wish to continue feeding a growing world population, the production of food will also need to be improved. The pressure that food production already places on the planet is no longer sustainable. In this respect, the Paris Climate Agreement and the United Nation's Sustainable Development Goals (or SDGs) must serve as a guide, says the EAT-Lancet report. To achieve these ambitious targets, it will be necessary for everyone everywhere to switch to a largely plant-based diet, whilst at the same time drastically reducing food loss and waste, and significantly increasing food production as a whole. So how can we do this?

First and foremost, it is necessary to make healthy food more readily accessible. This will require investment in marketing, information, education, food guidelines, etc. The amount of food derived from animals must be substantially reduced.

A major transformation will also be necessary in agriculture and fishing. The emphasis will need to be shifted away from the production of calories and towards the production of a varied diet that will benefit the health of both people and the planet. For example, this means that greater biodiversity should be encouraged, instead of the limited number of crops that are currently grown, primarily for use as animal fodder.

The report claims that we need a new agricultural revolution: yields must be increased, but in a sustainable manner, with greater attention to the more efficient use of fertilisers, water, nitrogen, phosphorus, etc. By 2040, it is essential that agriculture absorbs more carbon than it emits.

This presupposes that the world's population will continue to be fed with the produce of the existing area under cultivation. There must be no extension of the current amount of agricultural land and under no circumstances must there be further large-scale deforestation. The responsible management of the oceans is also a must.

Equally crucial is the battle to limit food loss on the production side and food waste on the consumption side. A 50% reduction is a minimum target.

Consent and resistance

Put simply, the bar has been set high. As the EAT-Lancet report makes clear, food will play a pivotal role in the 21st century. We need to seize the opportunity to develop a new food system that is focused on improving human health and supporting a sustainable environment.

The conclusions of the report have been widely accepted and are generally in keeping with the consumption trends that we can already see today, such as the growing popularity of plant-based eating patterns, the increased interest in local foodstuffs, an escalating aversion to sugar and additives, the rise in the use of sustainable farming methods, the growth of support for fair trade, etc. However, all these trends will need to be taken much, much further.

Perhaps for this reason, the recommendations have also provoked resistance. The agricultural sector feels itself unfairly targeted in discussions about surpluses of manure, use of water and the application of nitrogen, as a result of which it tends to react defensively. Consumers feel uncomfortable when their culinary traditions are brought into question. Are the scientists saying that we must abandon our favourite national dishes, such as British fish and chips and the American hamburger? Do we all have to start eating lentils and drinking oat milk? Or, worse still, develop an appetite for seaweed and insects?

However, the EAT-Lancet report also immediately attracted significant scientific criticism. Some see the document as promoting a one-sided vision that is not based on serious and well-founded scientific research. Clinical studies have not been able to prove conclusively that a vegan or vegetarian diet improves health and helps to keep illness at bay. Nor is there any firm evidence that red meat causes certain diseases, including cancer.

The largely plant-based pattern of eating that the report favours would also, in the opinion of some, result in nutritional deficiencies and is certainly not suitable for babies, children and pregnant women. What's more, for many people in poorer parts of the world the recommendations are simply unaffordable, say the critics.

For others, the very idea of a 'healthy planetary diet' is difficult to swallow, since the concrete circumstances on the ground are very different in different parts of the world. You cannot, they argue, draw up the same guidelines for the rich industrialised nations and for the poorer countries in the South.

Finally, some critics have insinuated that the report was compiled to serve the interests of the food industry. The EAT-Lancet commission was supported by FReSH (Food Reform for Sustainability and Health), a coalition of thirty or so multinationals, including major food producers like Danone, Kellogg's, Nestlé, PepsiCo and Unilever, as well as pharmaceutical companies, biotech players, chemical giants and major suppliers to the agricultural sector. The doubters fear a quid pro quo for companies and corporations that only see sustainability as a marketing tool.

From farm to fork

Europe is doing its best to play a key role in these developments. In 2020, the EU launched the 'From Farm to Fork' strategy as the central element of the European Green Deal, through which Europe hopes to become the world's first climate-neutral continent by 2050. There is much to be done and Europe bears a heavy responsibility: the EU is the largest importer and exporter of agri-food products and the largest market for fish and shellfish in the world. True, the European agricultural sector is the only one that has been able to cut its emission of greenhouse gases by 20% since 1990, but it is not enough. Even greater efforts still need to be made. With this in mind, the European Commission now aims to make the entire food chain from farm to fork more sustainable. Some 40% of agricultural subsidies during the next seven years will be devoted to the environment and climate. By

2030, a quarter of all agricultural land must be used for organic cultivation. The use of pesticides, herbicides and insecticides must be reduced by 50%. The Commission has opted to strengthen regional and local food systems, in order to create shorter food chains and reduce Europe's dependence on food transport over long distances. Other policy measures must lead to greater income security for farmers and a good earning model for the production of sustainable food.

The Green Deal wants to create an environment in which making the healthy and sustainable choice (read 'more plant-based choice') is the easiest choice. A framework for labelling will be developed, which, in addition to nutritional value, will also list climate, environmental and social effects. Clear labels are the best way to simplify and stimulate consumer choice for sustainable food. Europe has also declared war on food loss and waste, which it hopes to cut by half by 2030.

It is further planned to publish an EU code of behaviour for business and marketing practices. Here are some of the more noteworthy passages in the text. Food companies must ensure that 'their pricing policy does not undermine the citizens' perception of the value of food'. More concretely: 'Marketing campaigns in which meat is offered at very low prices must be avoided'. To help consumers eat more healthily, the Commission will also seek 'to restrict the promotion (through nutritional or health claims) of foodstuffs with a high fat, sugar and salt content'. The Commission is also considering the use of preferential VAT tariffs to stimulate the sale of organic fruit and vegetables.

'People like food that comes from nearby. That is what we have to give them,' said European Commissioner Frans Timmermans in a podcast on Belgian radio (Holderbeke, 2021): 'The label must contain information, in a manner that is easy to understand for everyone, about how the food was produced and details of its nutritional value. It must be made clear to people that eating more meat puts a heavier strain on the environment. It is our task to keep our citizens as well informed as possible, so that they can make informed choices. It is not our task to tell people what to eat.'

What does this new European policy have to offer for farmers? According to the European Commission, more sustainable business models, together with labelling and marketing standards that link production methods to consumer demand, will lead to higher revenues for food producers. In this way, farmers will be given a stronger position in the food supply chain. New opportunities for agriculture will also be created; for example, the production of vegetable proteins or in the bioeconomy. Technological and digital advances – such as precision agriculture – will result in higher yields and lower costs. By attempting to satisfy the increased demand for sustainable food, the agricultural sector will develop closer ties with consumers. Labelling and marketing initiatives must increase the export opportunities for European agricultural products.

'We must break with the practice of making agriculture ever more intensive,' says Timmermans. 'The European farmer must be retrained to become the guardian of the planet. But there must be a quid pro quo. We will need to find other sources of income for our farmers. If we pay farmers to put food on our plates, we should also pay them to protect nature. This will increasingly become the new business model for agriculture. Science will also need to be put at the service of the countryside and the farming community. Precision agriculture, the use of the latest technology, the extension of the broadband network into rural areas: these are also things that can help farmers to work more efficiently. Moreover, there is a group of young farmers ready and willing to embrace these changes. They can see that it really is possible to achieve good rates of production for healthy food with the use of significantly less fertiliser and chemicals. In this respect, farming is not a sector like any other. It will always need the support and guidance of the government.'

There is one other detail that is worthy of attention: in order to achieve the ambitious objectives of the sustainable Green Deal, legislators and the food industry must not be afraid to look at the possibilities offered by genetically modified crops. That, at least, is the conclusion of a European Commission report. A re-evaluation of the risks, and the strict regulations to which they are subject, is urgently needed. The current legislation covering this controversial technology came into force as long ago as 2001 and is therefore now outdated. Today, it is unquestionably possible to improve some existing crops genetically, say leading researchers in this field. This is something very different from the 'GGOs' of the first generation. A round of consultations is planned, but agreement still seems a long way off. While the scientists claim that the EU's sustainability objectives are unachievable without genetically modified food production, opponents are convinced that the report reflects the lobbying work of the agricultural and biotechnology sectors. The questions currently on the table include whether or not modified food can be safely consumed – by animals and/or humans – and, if so, how it should be labelled. At the time of writing, no genetically modified products are available on the European market.

Searching for solutions

In contrast, few people dispute that there are some serious shortcomings in our current food system, which urgently need to be corrected.

In the following chapters we will look at the ways in which science and the food sector are actively working together to find various solutions that can lead to a healthier and more sustainable food system. For example, urban agriculture or vertical farming can bring the production of fruit and vegetables closer to the consumer, while also making better use of scarce space and increasing productivity. Similarly, the development of healthy and tasty alternative plant-based products can make us less dependent on intensive livestock farming. The industry is also seeking to make use of new and more sustainable ingredients for food production: insects, for example, but also algae and seaweed. A completely new approach is the 'cultivation' of meat, poultry and fish in bioreactors.

THE SUSTAINABLE SHOPPING BASKET OF THE WWF

With its project #Eat4Change, the WWF wishes to encourage Europe's young citizens to modify their eating patterns and to make food production more sustainable. The organisation argues in favour of a flexitarian (semi-vegetarian) diet with more fruit, vegetables, legumes, nuts, grain products and plant-based alternatives for meat and milk, such as tofu, tempeh, quorn and soya drinks. It further recommends eating less meat (just twice per week) and fewer dairy products (three times per week), while also drastically reducing the intake of processed products like soft drinks, snacks and alcohol.

WWF advises young consumers to opt for products derived from sustainable production systems (organic agriculture, extensive livestock farming), which guarantee a fair reward to the farmer. Compared with our current patterns of consumption, this kind of 'sustainable shopping basket' could result in a 9% saving on the average weekly grocery budget. It would also require 37% less land, so that the pressure on nature could be alleviated. Similarly, the negative impact of farming on climate change would be halved.

The smart agricultural revolution

More efficient and more sustainable farming

How can we continue to feed the growing world population? How can we get the necessary food to the places where people live, which by 2050 will be predominantly in urban areas? Since these developments will require food production to increase by an estimated 70%, the answer has to begin with more efficient and more sustainable farming. But is it really possible to further increase yields, whilst at the same time reducing the impact of agriculture on the planet? Yields have already risen dramatically in recent years, thanks to significant upscaling and technological improvements. However, the negative consequences for the environment were considerable. Farming needs to become 'climate intelligent' and precision agriculture is part of the solution. Technology can help to boost yields, while at the same time reducing the use of resources such as water and fertiliser. In short, agriculture needs to get smart. We can expect the use of micro-sensors, drones with cameras, robotised weed and pest control, automatic harvesting machines, self-driving tractors, etc. The EU is already funding 'Robs4Crops' research to evaluate the potential for agro-bots to carry out repetitive and unpleasant tasks for which farmers find it difficult to recruit labour. Data analysis and artificial intelligence will make possible more efficient, more sustainable and more profitable agricultural practices. Vertical farming is also part of the plan. By working upwards (or downwards), farmers can increase the volume of space but not the surface area that they are able to cultivate. In other words, more crops but not more land. This type of agriculture is set to undergo a rapid transformation from hyper-local to large-scale industrial.

How agriculture will become high-tech:

- Tele-detection: the use of drones and satellites to monitor fields and livestock from a distance.
- Sensors and smart camera networks for guaranteeing the security of agricultural land and infrastructure.
- Software for the management, organisation and optimisation of all tasks on the farm.
- Robots for the replacement of manual labour for difficult, dangerous or expensive tasks.
- Precise agriculture and precise irrigation, supported by automation and robotisation.
- The development of new recipes for animal feed; for example, based on insects or seaweed.
- The development of new recipes for fertiliser; again, based on insects or seaweed.
- The development of biopesticides: environmentally-friendly pest control solutions based on living micro-organisms, such as bacteria, algae, viruses, fungi, pheromones, insects and plant extracts.
- Urban agriculture and new-generation agriculture (often in covered and strictly monitored spaces).
- Genetic modification, in which bioscientists alter the DNA of an organism; this allows them, for example, to make them more resistant to pests and diseases, or to make them more suitable for cultivation in dry regions, or to increase their storage life.
- Business-to-business e-commerce marketplaces for farmers, with products ranging from seeds to heavy equipment.
- ...

MORE STRAWBERRIES THANKS TO ARTIFICIAL INTELLIGENCE

The technology of the fourth industrial revolution, like AI, can lead to huge production increases in agriculture. An example? In China, four technology teams engaged in a four-month long competition with farmers to grow strawberries. The scientists won by a distance: on average they produced 196% more strawberries in terms of weight than the traditional farmers. The tech-wizards also outperformed the farmers in terms of return on investment, by an average of 75.5%. However, the research results do not confirm whether the strawberries actually tasted better or worse. During the experiment, the technology teams had the advantage of being able to adjust the levels of temperature and humidity through the automation of the greenhouses they were using. Intelligent sensors also made possible the more effective control of the use of water and nutrients, whereas the farmers had to perform these tasks by hand and based on their own experience. This competition was organised by Pinduoduo, China's largest agro-technology platform, and the China Agricultural University, with the Food and Agriculture Organisation of the United Nations as technical adviser.

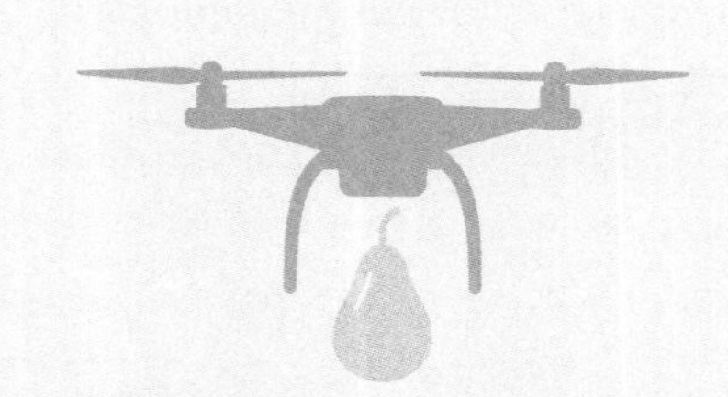

DRONES REPLACE FRUIT-PICKERS

The Israeli company Tevel Aerobotics Technologies has developed an autonomous robot that makes use of artificial intelligence to identify and pick fruit. The robot can work 24 hours a day and only picks ripe fruit. This innovation was a direct response to a perceived shortage of manual labour. 'There are never enough hands available to pick the fruit at the right moment and the right price. As a result, fruit was left rotting in the orchards and was sold for only a fraction of its peak value, causing the farmers to lose billions of dollars each year,' says the company. The robot uses AI to locate the trees, find the fruit amongst the leaves and identify its size and degree of ripeness. The robot then assesses the best way to approach the fruit in a stable manner, so that it can extend its plucking arm to harvest the fruit. The drones can clear an orchard without getting in each other's way, thanks to a single autonomous digital brain installed in a unit on the ground.

A smaller step from farm to fork

According to the United Nations Food and Agriculture Organisation, we are currently using about one-third of the world's total land area for livestock farming and crop cultivation (FAO, 2018). In other words, there is not much space for further expansion. If we wish to preserve biodiversity, we must avoid further deforestation and ensure that no new land is devoted to agriculture. But it will not be easy. To make matters worse, climate change is threatening the future use of part of the land currently available. Scarcity of water looks set to become a major problem. Some regions are already only suitable for very specific types of cultivation. For example, large parts of northern Africa are only suitable for growing olive trees. Moreover, the distribution of land for possible expansion is very uneven. There are still some possibilities in South America and sub-Saharan Africa, but almost none in South Asia, the Near East and North Africa. Unsurprisingly, the situation is even more acute in the densely populated industrial nations, where levels of demand are also the greatest.

For this reason, the idea of developing urban agriculture and vertical farming is not as futuristic as it might have sounded as little as a decade ago. Both concepts make possible the more efficient use of the scarce available space, whilst at the same time bringing food products closer to the consumers, as well as reducing both transport requirements and food loss. Innovative technology facilitates the growth of crops in these new environments all year round, under modified LED-lighting and in strictly controlled conditions. This not only increases yields, but also ensures the safety, the nutritional value and even the flavour of the harvest. It is also a more sustainable form of cultivation, requiring less use of water (thanks to its recycling), fertilisers and pesticides.

Is this all as positive as it sounds? That remains to be seen. According to the FAO, vertical farming and urban agriculture (broadly interpreted) currently represent some 15% of total agricultural production, but this primarily consists of leaf vegetables and herbs, which can easily be grown indoors. The real challenge will be to try and increase the diversity of the range of crops, many of which require too much space to be cultivated on an urban farm.

THREE TECHNOLOGIES, ONE FUTURE

There is quite a lot of verbal confusion in the world of vertical farming. Three English terms ending in '-ponics' are often used interchangeably – and therefore incorrectly. In reality, there are important differences between these three ponics: aquaponics, hydroponics and aeroponics. These differences relate primarily to the manner in which nutrients are given to the plants.

Hydroponics is the method that is most commonly used today in 'traditional' urban agriculture. The roots of the plants do not grow in full ground, but on a substrate – like rock wool or hydrogranules – and are computer-fed with a water solution that contains all the necessary nutrients. Most vertical farming concepts make use of this method.

Aeroponics does not make use of a substrate. Instead, the roots of the plants hang in the air and are moistened with a spray that contains all the necessary nutrients. The main advantage of this method is that the plants can absorb the nutrients more effectively and therefore grow more quickly. This is the method that NASA uses in the International Space Station (ISS). It is not only suitable for the growth of herbs and leaf vegetables, but also for fruits (strawberries, tomatoes) and root vegetables (radishes, carrots and even potatoes). The British start-up LettUs Grow is testing this concept in the form of 'hanging gardens' for the Waitrose supermarket chain.

Aquaponics is a variant of hydroponics, which combines fish cultivation and plant cultivation in a closed circuit. The excreta and food remnants of the fish serve as nutrients for the plants. In the reverse direction, 'wastage' from the cultivated plants can serve as food for the fish. Bacteria convert the fish waste into nutrients. As a result, it is not necessary to replace the water, which makes this a highly sustainable and waste-free concept.

Rooftop farming

Urban agriculture takes many different forms. One of the possibilities is growing crops in the open air or under glass on the roofs of buildings in towns and cities. This approach does not differ so greatly from traditional horticulture, but it has the big advantage of making use of space that would otherwise go unused in densely populated areas. One excellent example is the Belgian start-up Peas & Love, which sets up urban farms in cities like Brussels and Paris. The company primarily targets private individuals, who can rent 'gardens' measuring 3m², complete with a plant trough and a vertical plant wall. Peas & Love plants some sixty different crops, ranging from herbs to pumpkins and raspberries. To make things even better, they also do all the maintenance: all the customer needs to do is the harvesting (De Schamphelaere, 2019).

Another similar project is Roof Food, a Ghent-based co-operative that seeks to bring agriculture into the city. This initiative started with the installation of a vegetable garden on the roof of the city's De Punt business centre, the produce from which was later turned into some excellent dishes in collaboration with some of the best local chefs. The company is now mainly engaged in offering support to building promoters, the care sector, retailers and local authorities for the installation of urban farming sites and rooftop vegetable gardens.

Equally impressive is the urban farm in the Prinzessinnengarten in the Berlin district of Kreuzberg. The farm has an area of 6,000m² and grows a range of forgotten vegetable and herbs, as well as flowers. A series of hives with a population of some 10,000 bees ensures the necessary pollination. Visitors can eat meals that make use of the farm's own produce. In 2020, it was decided to split this initiative and a new 7.5 hectare community farm known as the Prinzessinnengarten Kollektiv was opened in the city's Neukölln district.

From the roof to the store

One of the most striking evolutions of recent times is that when project developers design housing developments for inner city areas, they are increasingly including the provision of rooftop greenhouses in their plans. In other words, homes and urban farming combined. You cannot bring fresh vegetables any closer to the consumer than that! In American cities like Westbrook, Philadelphia and Chicago a number of concrete projects are currently in the pipeline. In addition to fifty or so affordable housing units, there is also a vertical greenhouse with 6,500 m² of space, which produces an annual yield of 450,000 kilograms of fresh vegetables and has created more than 50 new jobs (Peters, 2021).

The supermarkets can also see the potential of this kind of rooftop horticulture. In Belgium, the Delhaize chain opened a vegetable garden on the roof of its Boondael supermarket in Elsene in 2017. The retailer grows different kinds of lettuce, but also tomatoes, aubergines and courgettes, all of which are sold exclusively in the supermarket. The main aim of launching this initiative was to study the possibilities that urban agriculture has to offer, but Delhaize has also added a social dimension by linking it to an educational programme (Van Rompaey, 2017).

The Elsene project was a European first, but not a world first. The American Whole Foods Market chain (now part of Amazon) opened a greenhouse and vegetable garden on the roof of its branch in Brooklyn as long ago as 2014. Do projects of this kind have a future? For the time being, that remains an open question. The projects mentioned above were designed first and foremost to test the viability of the concept and to raise public awareness. As an added bonus, it was also good public relations.

Thirty metres under the ground

Urban agriculture can also be practised indoors or even under the ground. The latter option requires the use of some very advanced technology. One of the pioneers in this field is Farm.One. The company was set up in New York in 2016 and its vision was to grow rare, interesting and unusual products for chefs, making use of the new technology of vertical farming. In addition to its large flagship farms, the company also sees potential in mini-farms for schools, companies, restaurants and supermarkets like Whole Foods (see above) and Eataly. Individual customers can also take out a subscription, which entitles them to three recyclable boxes of leaf vegetables, herbs and flowers each week, selected from a range of almost 700 varieties. This is enough for six king-sized salads, with a few herbs left over for the garnishing. Farm.One sees it as its calling to make the public better informed and better aware about the possibilities and the advantages of vertical farming.

In the cellars of the former Belle-Vue brewery in Sint-Jans-Molenbeek, the Brussels start-up MicroFlavours grows micro-vegetables on an area of some 400 m^2. Each day, they harvest between 10 and 15 kilograms of vegetable sprouts, packed full of flavour and nutritional goodness. These are then delivered to 70 top restaurants in the Belgian capital, as well as to a number of local supermarkets. RetailDetail is bringing the concept to Antwerp, where it will be tested in the Foster 'living retail' laboratory.

But one of the most adventurous projects of this kind has undoubtedly been launched by Growing Underground, which makes use of disused tunnels 30 metres under the ground in the London district of Clapham to cultivate micro-vegetables and salad products under LED-lighting in a pesticide-free environment. Its customers include such famous names as Waitrose, Marks & Spencer, Ocado and Whole Foods Market. In just four hours, the harvested products can be on the chopping board in the customer's kitchen. You cannot get any fresher than that!

High-tech herb cultivation

Retailers are always quick to see the benefits of technology. Consider, for example, the following interesting project set up by the Colruyt Group. In a former distribution centre near the company's headquarters in Halle, they are growing basil under ultra-violet light in a moist atmosphere at a temperature of around 30°C. This is state-of-the-art technology, in which every detail needs to be perfect. Each day, the plants are given a carefully calculated cocktail of light, air and water. This strengthens the basil, while the right combination of red and blue light ensures that it grows twice as fast and contains more aromas. As the icing on the cake, it can also be kept for longer, which leads to less food loss. Colruyt developed this technology in-house, with the assistance of the Catholic University of Leuven for the fine-tuning of the LED-lighting (Van Rompaey, 2020).

Why is the Colruyt Group doing this? 'Sustainability is in our DNA, and that is certainly the case for our Bio-Planet project. We are looking for answers to well-known societal problems: the growing world population, scarcity of water, the shortage of agricultural land and raw materials ... We need to think carefully about the food of the future. Our system for basil cultivation is sustainable and local: it allows us to grow more with less space. This reduces the size of our ecological footprint,' explains Fabrice Gobbato, the director-general of Colruyt's organic chain Bio-Planet. Vertical farming needs twenty times less area, 90% less water, 50% less nutrients and no pesticides. The concept works entirely on green energy and is highly energy efficient. And because the herbs are grown locally, close to one of the group's new distribution centres, it requires five times fewer kilometres to transport it.

Greenhouses in the supermarkets

What Colruyt does on a relatively large scale can also be done on a small scale. This is the speciality of the German InFarm scale-up, the European leader in this field. After a pilot project in Berlin, the company developed a modular concept of in-store greenhouses for the food service wholesalers of the Metro group. The first of these high-tech supermarket greenhouses outside of Germany, measuring 25 m^2, was installed in a Metro store in Antwerp. Again, the cultivation takes place under adapted LED-lighting, which makes the resulting herbs tastier and more nutritious. Local chefs are also able to find less common varieties and can even make suggestions about what they would like to see grown. This partnership between InFarm and Metro is still active. The largest in-store greenhouse in Europe is currently located in the Metro supermarket in the French city of Nanterre and has a surface area of 80 m^2.

The plants are cultivated on water beds that are enriched with nutrients, such as calcium, potassium and magnesium. The water is constantly recycled in a closed circuit, so that nothing is wasted. Each greenhouse has its own robot and computer, which is connected to the internet and to twenty or so sensors that measure different parameters, including water, light intensity, ambient temperature, etc. The water level can be adjusted by activating pumps and it is also possible to reproduce a night and day cycle. Everything can be monitored and controlled at a distance and nothing is left to chance. According to InFarm, these mini-greenhouses are much more efficient and ecological than traditional agriculture: the system uses 95% less water, 90% less transport and 75% less fertiliser.

In the meantime, InFarm is now also collaborating with other retailers, including Marks & Spencer in the United Kingdom. In 2019, M&S launched a project with mini-greenhouses in a number of its stores in the London area. Customers were able to buy basil, mint, parsley and coriander grown on site. The Selfridges chain of department stores has also had a greenhouse installed by InFarm in its branch in London's Oxford Street, as part of the company's five year plan entitled 'Project Earth'. In the Danish capital Copenhagen, the German scale-up is working with Irma, a daughter company of Coop Danmak. In the United States, Kroger was the first out of traps, with a pilot store in Seattle. In France, the Intermarché group also has greenhouses in a number of its supermarkets.

You can watch a short film about the InFarm mini-greenhouses in Marks & Spencer stores here:

https://youtu.be/eT7ZXO1IPS4

Container farms

A different approach to the same problem has been adopted by the Israeli company Vertical Fields. In this case, the plants are grown in containers, but placed against the walls, rather than in troughs. Other than that, the principles involved are very similar. Sensors monitor parameters such as temperature, humidity, lighting and nutrition. An algorithm makes corrections, when necessary. The result is a higher yield with better quality, without the use of pesticides. The focus is on herbs and leaf vegetables, like lettuce, kale, spinach and pak choi. Vertical Fields sees further potential for the use of containers on supermarket car parks, as well as in hotels, restaurants, hospitals and universities.

Other players opt for a more industrial approach. One example is the massive indoor farms installed in the greenhouses 3.0 of the Belgian scale-up Urban Crop Solutions, which aims to export large-scale vertical farming factories to the world. Cultivation takes place on a substrate in beds of 250m² at eight different levels. Algorithms control the fertilisation, humidity, oxygen and CO_2 levels, and the lighting. The plants are completely isolated from the vagaries of the weather and damage by insects. Growth continues throughout the year and use of water is limited, so that this type of indoor farm can even be installed in a desert. Urban Crop Solutions cultivates more than two hundred different crops, from lettuce and spinach to herbs and paprika.

The company does not farm itself. Instead, it sells its technology – often in containers – to interested players worldwide (De Schamphelaere, 2019). These include IKEA: the giant furniture retailer is testing the cultivation of leaf vegetables in UCS containers alongside its stores in Helsingborg and Malmo. The plants are nourished with food waste from the store restaurants. The resultant yield is between 15 and 20 kilograms per day. The salad products are intended in the first instance for the staff restaurant, but it is planned to later extend this to the public restaurants. 'The aim is to create a closed and circular food chain. We are looking forward to seeing the results of this interesting test,' says Catarina Englund, the global sustainability innovation manager at IKEA.

FARMING FISH ON THE MOON?

It even seems probable that at some point in the not-too-distant future aquaponics can help to provide a varied diet for astronauts in space and colonists on the moon. Researchers at the French Institute for Research into the Exploitation of the Sea (IFREMER) and the University of Montpellier are working on a project called 'Lunar Hatch', which aims to make possible the cultivation of fish on the lunar surface. The idea is to send pre-fertilised fish eggs into space, where they can be converted into fish by the available water on the moon. 'I suggested sending eggs rather than fish, because eggs and embryos are very strong,' said marine biologist Cyrille Przybyla in an interview with Hakai, an online scientific magazine. The first tests do indeed suggest that the eggs can survive well in extreme conditions. In particular, sea bass is a good candidate for further trials. For potential colonists, the farming of animals and plants on the moon would not only be important from a nutritional perspective, but also from a psychological one, as a reminder of life on Earth.

Tomatoes and fish, a perfect match

Or perhaps aquaculture or aquaponics (see inset) are the future for urban agriculture. This remarkable concept combines plant cultivation with the cultivation of fresh water fish in a circular system. The fish and the plants share the same water; algae in the water form the main source of food for the fish and bacteria convert the excreta of the fish into nutrients for the plants. This type of aquaculture is currently increasing all around the world. Species of fish cultivated in this manner include catfish, carp, tilapia and salmon, but shrimps can easily be farmed using the same technique.

A good example in Belgium is Aqua4C, based in Kruishoutem. Since 2014, they have been farming jade perch, a fish that grows in pure rainwater and lives on a vegetarian diet. The fish farm works closely with its neighbour, Tomato Masters. Their combined system is ingenious: residual heat from the tomato greenhouses heats the water in the tanks where the jade perch swim. In return, Aqua4C provides nutrient-rich waste water for the tomatoes. Thanks to this unique collaboration, the fish are cultivated in a quasi-closed circular system.

According to the producer, the jade perch is the most sustainable fish in the world. It is a 100% vegetarian, so that it requires no fishmeal or fish oil. This specific diet results in the fish having high concentrations of beneficial omega-3 fatty acids (hence its Belgian name: omega perch). Nor is it treated with antibiotics. Jade perch is already on sale in Carrefour, Spar, Albert Heijn, Bio-Planet and specialised fishmongers.

One of Europe's largest urban aquaponics farms was opened in 2018 on the roof of the Foodmet abattoir in the Anderlecht district of Brussels. The 4,000 m^2 abattoir farm combines hydroculture with aquaculture to create an intelligent circuit in which water, CO_2 and the excreta of the fish are all collected and recycled. The project is the brainchild of the Brussels start-up BIGH (Building Integrated Greenhouses). The company hopes to produce 35 tons of striped bass and 15 tons of tomatoes each year, as well as 2,700 pots of organic herbs each week.

The Stadt (City) Farm in Berlin also produces fish, seasonable vegetables and tropical fruit in a highly concentrated urban area. They claim to be Europe's largest

AquaTerraPonik urban farm under glass, situated in the Herzberge Landscape Park. It has an annual production of 50 tons of African catfish and 30 tons of lettuce, herbs, tomatoes and cucumbers, but also more exotic products like bananas, Ceylon spinach and passion fruit. The cultivation takes place in a closed circuit. Excreta from the fish is converted by bacteria into nutrients. The water flows through soil-filled plant basins, in which worms are also living. The plants absorb the nutrients and so purify the water, which is then returned to the fish. And all in the heart of the city. In short, a project that gives you a taste for more!

Take a look at a film about the aquaponics project of the Brussels start-up BIGH:

https://youtu.be/JfVhrnerzqo

NO SUNLIGHT NEEDED

A spectacular example of indoor agriculture is the Super Sprout Factory of the Murakami Farm Co. It is the largest artificially-lit plant factory in Japan and its purpose is to increase the production of broccoli sprouts, which are much in demand because of their remarkable beneficial properties. Broccoli sprouts contain high concentrations of sulforaphane, an active compound that can help to combat cancer, cholesterol, diabetes, skin damage, autism and so much more. In this high-tech factory the cultivation of the sprouts takes place in huge rotating cisterns and is controlled by a central system that is largely automated.

Have a look at this video about the 'Super Sprout Factory':

https://youtu.be/F8J6HYDODvk

HOW AGRICULTURE AND HORTICULTURE CAN BE MADE MORE SUSTAINABLE

If you want to take steps towards achieving greater sustainability, this does not per se mean that you need to practise urban farming on the roof of a supermarket or in a disused railway tunnel under the ground. The traditional agricultural and horticultural sectors are also aware of the need to cultivate more sustainably and are developing some interesting initiatives.

Some examples? Super-insulated greenhouses that make use of condensation pumps have the potential to become climate neutral over time. New types of LED-lighting are capable of saving 37% on electricity use. Reusing rainwater over and over again can ensure that not a single drop is lost. In livestock farming, one of the key aims is to eliminate the use of imported animal feed – mainly soya – by allowing cows to graze more on pastureland. In this way, the cattle can convert vegetable proteins that humans cannot digest into proteins that are digestible. It is also possible to feed livestock with residual products from industry, such as beet pulp from the sugar industry or beer draff (a malt residue) from the brewing industry. Cows not only digest these alternative foodstuffs better but they also produce less methane gas as a result, which benefits the climate – and therefore us all.

HOW REGENERATIVE AGRICULTURE WANTS TO HEAL THE PLANET

Today, it is no longer enough simply to reduce the ecological footprint of food production. The question is how we can replenish nature's limited natural resources instead of further exhausting them. The starting point for this process is what is now generally referred to as regenerative farming. This type of farming aims first and foremost to improve the quality of the soil and subsoil naturally, without the use of chemicals and fertilisers, as well as to increase biodiversity. This is coupled to sustainable energy-saving measures and the reuse of (rain)water, with the aim of becoming climate neutral by absorbing more CO_2 than is emitted. In short, regenerative farming goes much further than organic farming.

The problem is, of course, that during the previous century the agricultural sector invested heavily in intensification and industrialisation. Monoculture has become the norm, whereas it is multi-crop cultivation that offers the greatest ecological benefits. As a result, farmers need to use fertilisers and chemicals to prevent their soil from becoming exhausted and to fight pests and diseases. This is clearly unsustainable in the long term.

But how do you change? That is by no means self-evident. On the bright side, studies have shown that regenerative farming has the potential to be 78% more profitable than traditional farming: yields are lower, but so are the costs (Lacanne, 2018). In other words, there is hope. Worldwide, multinationals like Danone, Kellogg's, General Mills and even the Kering luxury group are supporting farmers to make the switch to this new kind of farming. Similar regenerative models are being developed for factories. There is still a long way to go, but at least we are making a start.

THE PROBLEM WITH MONOCULTURE

The worldwide upscaling and industrialisation of agriculture has improved efficiency enormously, but it has also led to the impoverishment of the range of crops produced. Of the roughly 30,000 species of crops that are potentially suitable for human consumption, modern agriculture makes use of less than 0.1%. According to the United Nations Food and Agriculture Organisation, maize, rice, wheat and potatoes account for almost half of the world harvest of vegetable food. Of the 40 or so species of domesticated livestock, just five of them provide by far the largest proportion of our animal foodstuffs (meat, milk, eggs).

Three-quarters of the food that is consumed by humans worldwide is derived from a paltry twelve varieties of crops (sugar cane, maize, rice, wheat, potatoes, soya beans, sugar beet, palm nuts, tomatoes, barley and bananas) and five species of animals (cattle, chickens, goats, pigs and sheep).

This is reflected in what you can see on supermarket shelves. Of the more than 400 edible varieties of bananas, you can find just one in our stores: the Cavendish. And that is a problem, because this plant is now being threatened by a new variant of fungus that causes the so-called Panama disease, which has spread from Asia to also infect plantations in Africa and South America, which are the most important banana producing regions. Because these plantations are based on a monoculture that makes use of genetically identical plants, the fungus can infect them all at lightning speed. One possible solution is to diversify this genetic uniformity by planting different varieties of banana alongside each other. Biodiversity protects our food production, as well as offering us greater variety on our plates, in keeping with the principle of 'use it or lose it'.

In this respect, a countermovement to uniformity is starting to emerge. For example, in recent years there has been increasing interest in older and more authentic species of cattle. Much the same is true for 'forgotten' vegetables, which are once again finding favour, not only in the kitchens of top chefs but also in the supermarkets. Parsnips, root parsley, kale, beets, cavolo nero, scorzonera and turnip never disappeared completely, but are now back with a vengeance. Even consumers are finding their way back to essentially 'old' concepts like 'the edible landscape' or 'the food forest': ecosystems that are self-sustaining, in which you do not need to hoe or use fertiliser, or even water. This is now generally referred to as permanent agriculture or permaculture. However, these are often small-scale projects; it remains to be seen whether the concept is capable of upscaling and commercialisation.

The plant-based revolution

One giant leap for mankind ...

Probably the most striking and the most dominant food trend of recent years has been the rapid increase in the attention devoted to more plant-based patterns of consumption. In industrialised Western countries, we ate – and still eat – huge amounts of animal-based products. For many consumers, meat and dairy are a daily part of their diet, and one that they hardly ever stop to think about. But it hasn't always been that way. Until roughly a century ago, meat was a luxury product, reserved for the rich or for special celebrations, such as Christmas. As general prosperity gradually increased, meat eventually became affordable and available for everyone. Nowadays, shoppers in supermarkets are bombarded with special offers for mince, chicken fillets and salami. And for many of these shoppers eating meat and/or cheese for breakfast, lunch and dinner is the most normal thing in the world.

Or it was. The realisation that things cannot be allowed to carry on this way is slowly sinking in, not only amongst the wider population, but also in the business world. No less a person than Bill Gates, the co-founder of Microsoft, co-president of the Bill and Melinda Gates Foundation and chairman of the Breakthrough Energy Ventures investment fund, now believes that the rich countries must switch to an entirely plant-based diet. He even regards it as 'unavoidable'. It is only in the world's very poorest countries that this ambition will be difficult to achieve.

If people need to be weaned off their addiction to meat, what are the alternatives? Not all of these alternatives provide the same benefits in terms of sustainability. The biggest gains can probably be achieved if we replace meat with legumes for at least part of our protein intake. Legumes are a complete source of protein that require very little preparation before being ready for consumption. There are also gains – but significantly fewer – to be made if we replace meat with meat substitutes or insects, providing they are eaten in their entirety. Alternative products that require a considerable degree of industrial processing provide only limited sustainability gains, or even none at all. That is the case, for example, in the extraction of protein from insects and algae for the production of cultured meat (Van der Weele, 2019).

Making Western eating patterns more sustainable not only presents the food industry with huge technological challenges – think of the slow progress being made by start-ups in the cultured meat sector – but also demands major social and institutional change. There is already resistance to these changes; for example, from the agricultural sector and part of the traditional food industry. There are also important hurdles to be overcome in terms of the necessary regulatory measures. But the biggest question of all is this: will consumers be willing to give up their old ways of eating and switch to something completely new? Food is a very emotional subject, and one where old habits do indeed die hard. Getting people to eat more beans and plant-based burgers is probably feasible, but the step to cultivated meat, algae and insects currently seems like a bridge too far. In the next chapter, we will look at the different possibilities.

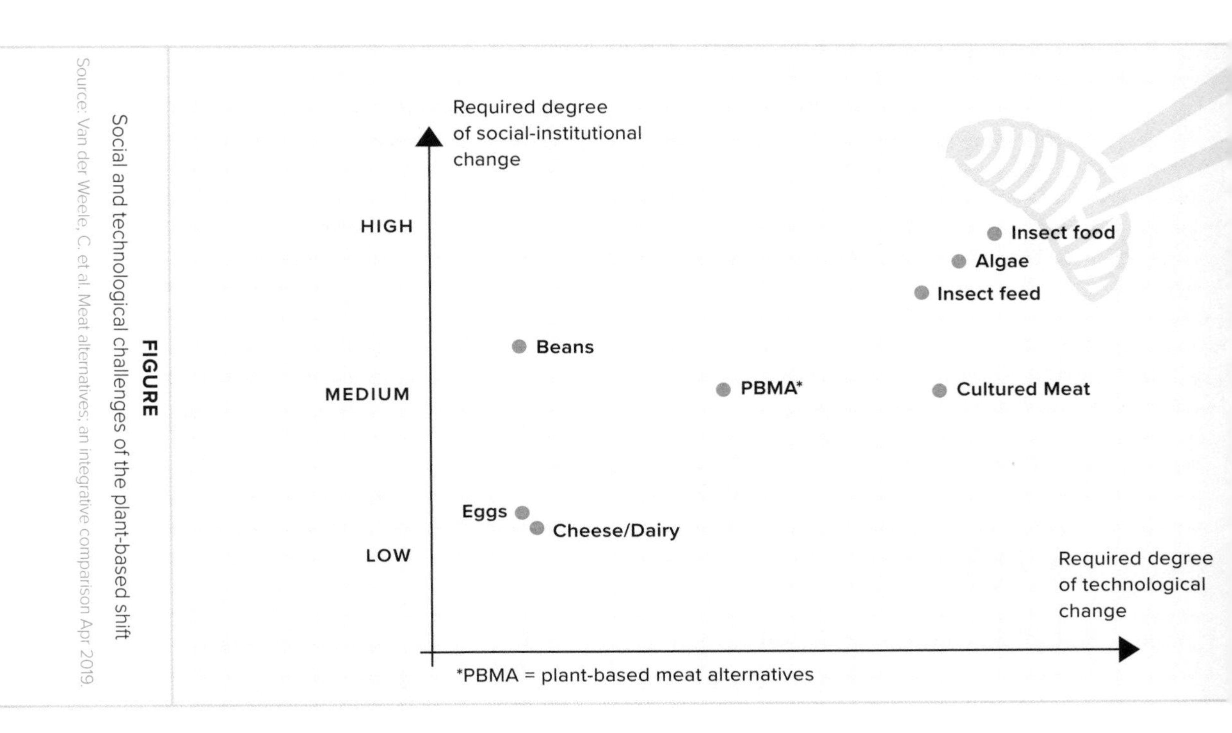

FIGURE
Social and technological challenges of the plant-based shift
Source: Van der Weele, C. et al. Meat alternatives: an integrative comparison Apr 2019.

The trend towards flexitarianism

The number of strict vegetarians and vegans remains limited. It is difficult to give accurate figures, since the statistics are usually based on fairly limited consumer research surveys. It is estimated that roughly 1% of the population in the Benelux is strictly vegan. This means that they eat no products of animal origin whatsoever: no meat or fish, no milk, eggs or cheese, not even honey. Strict vegans also refuse to wear leather or wool. The estimated figures for the number of vegetarians tend to vary. According to the EVA, a vegetarian interest group, their followers now represent some 9% of the Belgian population. Vegetarians eat dairy products and eggs, but not meat and fish. Both groups are larger in the major cities than elsewhere and both trends are particularly popular with younger people.

Having said that, there is a third new trend that is also attracting an increasing amount of support: flexitarianism. Flexitarians generally abstain from animal-based products for one or more days each week, replacing them with plant-based alternatives. Others prefer to replace specific animal products in their diet entirely; for example, they switch from dairy milk to plant-based milk, like oat milk or soy milk. It is thought that almost half of the population now regularly eats plant-based products on this kind of rotational basis. Variation, health, animal welfare and environmental friendliness are the most important motivating factors. Moreover, the range of such products has increased exponentially in recent years, as has their quality (flavour, texture) and the attractiveness of their price.

According to a study conducted by the Boston Consulting Group and the Blue Horizon vegetarian think tank, Europe and the US will reach 'Peak Meat' in 2025, the point at which meat consumption will subsequently start to decline, following decades of unbroken growth. In developing markets consumption will temporarily continue to rise, but the overall worldwide consumption of plant-based alternatives will rise correspondingly, amounting to an estimated 11% of the total protein market by 2035 – with the possibility to increase this figure to 22%, if the pace of the necessary technological and regulatory change can be speeded up (Witte, 2021). To convince consumers to 'play ball', it will be necessary to develop alternatives that are equivalent in terms of flavour, texture and price. As far as this latter aspect is concerned, it is expected that the price of meat substitutes will have fallen to the same level as 'real' meat by 2023. By 2035, the market for plant-based alternatives for meat, dairy products, eggs, fish and other seafood is likely to amount to a turnover of 240 billion euros.

MEAT SUBSTITUTES: THE MAIN PLAYERS

The term 'veggie burger' is often used as a synonym for meat substitute. Because we eat burgers in a bun with lashings of sauce, the flavour difference with real meat is often less noticeable. It is this fact that made burgers such a good entry category for the producers of meat alternatives. In the meantime, substitutes have been developed for many of the most popular meat products (meat balls, sausages, mince, schnitzels, etc.), but imitations also exist for chicken fillets, chicken nuggets, fish fingers and meat derivatives like ham and spam (luncheon meat).

One of the very first pioneering products in the vegetarian market was Quorn, which appeared as long ago as 1985. Quorn is made via the fermentation of mycoprotein, a natural and nutritional fungus that is a good source of protein and rich in fibre. The first generation product also contained added protein from eggs and was therefore vegetarian, rather than vegan. Quorn is now available in a vegan variant.

Garden Gourmet is the plant-based brand on which food giant Nestlé is placing its hopes for the future. The products are made on the basis of soya and wheat proteins. Their extensive range contains alternatives for all the most popular meat and poultry products.

Vivera, a Dutch brand, is the European number three in this sector. It has been making meat and fish substitutes since 1990, based on vegetables, soya protein and wheat protein. Originally, Vivera was part of the Encko meat processing group, but decided to go it alone in 2018, specialising exclusively in plant-based products. And with success: in 2021 the company was taken over by the Brazilian meat giant JBS.

Beyond Meat is a producer from the Silicon Valley generation, which has developed advanced methods to imitate the flavour and texture of meat as closely as possible. Pea proteins form the basis; beetroot juice gives the colour. Their products are soya-free, gluten-free and contain no genetically modified ingredients.

Impossible Foods is another high-tech food company. Its speciality is the 'bleeding burger'. The basis for the meat substitute is soya, but it also contains soya heme, which looks like real blood. Because this burger contains genetically modified ingredients, it cannot currently be sold in Europe.

The Vegetarian Butcher was founded in the Netherlands in 2010 by Jaap Korteweg. His ambition: to make plant-based meat the norm and reduce the amount of livestock farming. At the end of 2018, the company was taken over by Unilever. Its product range includes products in the style of beef, pork and chicken, based on soya.

Greenway started in 1997 as a vegetarian restaurant in the Belgian city of Ghent and has since grown into a producer of plant-based products that can be found in supermarkets and catering outlets throughout the country. The majority of their products do not contain soya or wheat.

Valess is the vegetarian brand of the FrieslandCampina dairy group. Its meat substitutes are based on milk and are therefore not vegan.

The battle for the vegetarian consumer

The food, catering and retail industries no longer see plant-based products as a marginal phenomenon, but rather as an interesting growth segment. In coffee bars, drinking a cappuccino made from oat milk is now seen as something normal. Burger chains now offer standard alternatives for their vegan customers: for its Rebel Whopper, Burger King joined forces with The Vegetarian Butcher, while McDonald's developed its own McPlant line and Ellis Gourmet Burger serves a Beyond Burger. Numerous major food multinationals are now fervent advocates of the plant-based revolution. It all means that the battle for vegan and vegetarian customers is becoming fiercer than ever.

You don't believe us? Just check it out. According to food producer Nestlé, veganism is one of the fastest-growing food trends alongside gluten-free and lactose-free. The company sees plant-based food developing over time into a business worth billions. With this in mind, Nestlé already sells vegan ice-cream (in the form of various varieties of its Häagen-Dazs brand) and is working on a vegetable milk based on nuts and berries. Under the Garden Gourmet label, the food giant is also commercialising a range of plant-based meat substitutes.

What's more, Nestlé is by no means the only multinational to see the huge profit potential in plant-based food. At the end of 2018, Unilever acquired The Vegetarian Butcher, a Dutch producer of meat substitutes, seeing it as a launching pad for conquering a worldwide market. The multinational had already launched a vegan Magnum ice-cream and also sees a big future for plant-based sauces (mayonnaise and others) in its Hellmann's range. Not surprisingly, the switch to plant-based products plays an important role in Unilever's new 'Future Foods' strategy.

'I think that we are still only at the very start of market growth for meat and dairy alternatives,' says Hanneke Faber, head of the food division at Unilever, in an interview with the *Financial Times*. 'Today, these alternatives form only a fraction of the total market for meat and dairy products. In most developed countries, the figure is around 5%, but some predictions say that this will eventually rise to 50% or more' (Evans, 2020). The company is setting its focus on meat substitutes, milk-free ice-creams and vegan mayonnaise without eggs, but has no ambitions in the milk substitutes market, where rival Danone leads the way. As far as its meat substitutes are concerned, Unilever sees soya as its most important ingredient, although algae are also set to play an increasing role in the years ahead.

Similarly, Upfield – the former (but now independent) margarine branch of Unilever – is determined to grow in the market for plant-based food. The company is developing 100% vegetable variants of its margarines under brand names like Becel, Bertolli or Solo, and with Violife has already launched a plant-based cheese alternative. This is currently a small market, but also a much underrated one, says Upfield, with potential for huge growth.

Dairy producer Danone also intends to stay ahead of the game in its sector. With this in mind, it acquired WhiteWave Foods back in 2017. WhiteWave is the mother company of Alpro, a leading manufacturer of plant-based dairy alternatives. What's more, the company is already marketing plant-based versions of well-known brands that consumers have only previously known as pure dairy brands. For example, since 2020 it has been selling a dairy-free version of its Actimel health drink. The same is true for its Danio and Danette ranges. 'Danone will increasingly become a flexitarian company,' said country manager Nathalie Pfaff of Danone Belux in an interview with RetailDetail. This is in keeping with their sustainable 'One Planet, One Health' vision, with which the company hopes to respond to modern consumer needs. 'We want to play an active role in this evolution: for us, dairy and dairy alternatives are complementary products.'

Retailers also believe in the potential of alternative products. For example, Marks & Spencer recently launched a new and ambitious vegan range with 60 different references under the brand name The Plant Kitchen. It includes ready-to-eat dishes such as pizza, potato salad, vegan mac & cheese or Thai curry, but also ingredients to make your own meals, like tofu and vegetarian sausages. 'This is the year when plant-based food will truly become mainstream,' says April Preston, head of product development at M&S. British food market leader Tesco is also tipping veganism as the hot culinary trend of the moment and rival Sainsbury's is also expanding its range. In the Benelux, it is retailers like Albert Heijn – the first supermarket to have a refrigerated vegan display – and Delhaize that are setting the pace (Van Rompaey, 2019). However, Lidl is not far behind. Even though this discounter generally has a limited range, it continues to systematically expand its offer of meat substitutes, of which roughly thirty are now available in its Belgian stores.

DAIRY SUBSTITUTES: THE MAIN PLAYERS

It all began with soya milk (which we are no longer allowed to call 'milk'!), but since then the range has diversified hugely, so that a wide variety of alternative dairy products are now available. There are substitute dairy drinks based on rice, almonds, cashew nuts, coconuts, oats and hemp. In addition, there are plenty of alternatives for dairy-based yoghurt, desserts, ice-cream and cheese. There is even a vegan egg. It is worth noting that the traditional dairy producers no longer attempt to ignore this segment of the market: they can now see its potential as an opportunity, rather than as a rival.

Alpro was one of the first pioneers in plant-based drinks. This Belgian company began producing soya drinks as long ago as 1979, long before dairy alternatives became fashionable. In 2009, the company, which was part of the Vandemoortele group, was taken over by the American WhiteWave Foods, which in turn was taken over in 2016 by Danone. Alpro products are now sold in more than 50 countries.

Another leading name on the supermarket shelves is Dream, a brand of the American Hain Celestial group, which specialises in health foods. The brand offers a varied range of drinks based on rice, soya, coconuts, almonds and cashew nuts, but also markets vegan chocolate, yoghurt alternatives and ice-creams.

Abbot Kinney's was founded in 2014 in Amsterdam, but is now owned by the Wessanen natural food group. The company makes alternative drinks, yoghurts and ice-creams based on coconuts and nuts.

Danone is not the only 'pure' dairy group that is now moving into the plant-based sector. The Swiss group Emmi, well known for its Caffè Latte coffee drinks, had developed its Beleaf brand, which markets alternatives for yoghurts, protein shakes and drinks. Similarly, the Danish multinational Arla Foods launched a new plant-based brand in 2020 under the name Jörd, which in the first instance will focus on oat drinks.

Oat drinks are currently hip and oat cappuccino is now a must-have for every coffee bar. The undisputed leader in this field is the Swedish Oatly company, which started in the 1990s as a producer of oat drinks for people who did not like (or could not tolerate) cow's milk, but today is supported by a galaxy of famous investors that include Oprah Winfrey, Natalie Portman and Jay-Z. Not surprisingly, perhaps, Oatly is now stock-listed. Proof, if any were needed, that plant-based is no longer an obscure niche. It has lost its 'woollen socks' image forever!

Growth opportunities

How big is this new market? And how big will it become? Between 2010 and 2020, the retail sale of meat and dairy alternatives in Europe has grown each year by almost 10%, which has resulted in a doubling of market size over the decade. In fact, the sale of meat substitutes has grown by as much as 121%, from 625 million euros in 2010 to 1.38 billion euros in 2020. During the same period, the turnover for plant-based dairy products grew from 1.5 billion euros to 3 billion euros.

Although some commentators claim that meat and dairy products will eventually be replaced entirely by alternatives, researchers at the ING Bank do not see this as a real possibility. At the present time, substitute products account for just 0.7 of the European meat market and 2.5% of the dairy market. Even if we assume a continuation of the current strong rate of growth (+/-10%), it would take until 2060 before the alternatives have displaced meat and dairy completely (Geijer, 2020). A more realistic assessment sees the market for meat and dairy substitutes increasing to 7.5 billion euros by 2025: 2.5 million euros for meat and 5 million euros for dairy. The European market share will also rise respectively to 1.3% and 4.1%.

Within the dairy segment, there is a huge difference between milk alternatives and cheese alternatives. Plant-based 'cheeses' are currently a premium niche product, which find it harder to reproduce the characteristics of animal-based cheeses. In contrast, plant-based milk now accounts for almost 10% of the total European milk market. In Belgium, the figure is as high as 14% and in the Netherlands it is 12%. The negative impact of the cheese figure means that the total category of dairy alternatives is good for 3% of total share in the dairy market.

What about meat? Although at the present time meat substitutes only account for just 1% of the total meat market worldwide, this figure has the potential to rise to 10% by 2029, according to a study by the British bank Barclays: in other words, ten times more than the current turnover, which amounts to some 12.5 billion euros. To make this prediction, the analysts compared market share evolutions for other alternative food categories, such as plant-based dairy substitutes (now good for more than 13% of the dairy market) or craft beer in the United States (good for 12% of the market). However, the market share figure for meat is unlikely ever to rise much higher than that without further technological progress: today's alternative products can replace mince, burgers and sausages, but not, for example, a fillet steak.

Be that as it may, it is clear that meat substitutes are appealing to a growing number of consumers, and not just to vegetarians and vegans. The fact that major players like Nestlé, Tyson, Conagra and Kerry are moving into this market speaks volumes, says the Barclay report. Even the big fast food chains are starting to see the potential. Following Burger King (which works with Impossible Foods in the US and The Vegetarian Butcher in Europe), McDonald's has now launched its own McPlant line of meat substitutes: in the first instance, a plant-based burger and alternatives for chicken and sausage rolls.

A BEAUTIFULLY MARBLED STEAK ... FROM A MACHINE

A beautifully marbled filet mignon made exclusively from plant-based ingredients? That is the impressive claim of the Juicy Marbles start-up. The company does not culture its meat in a laboratory or print it on a 3D printer. No, it has developed its own machine specifically for this purpose: the Meat-o-matic Reverse Grinder™ 9000.

This machine reproduces the muscular texture of meat by building up successive layers of soya and wheat proteins. It is a simple, 100% physical process that only makes use of natural ingredients. The quality of the protein is crucial for a nutritious steak; the quality of the texture is crucial for giving it good bite; but the real challenge is to recreate the intramuscular fat that makes a real marbled steak so juicy and so tasty. So how do they do it? The secret is unsaturated sunflower oil. Red beetroot juice provides the colour and natural flavourings do the rest.

In the first instance Juicy Marbles sells its plant-based steaks direct to consumers as part of a test programme, but there are plans to open a production facility at Austin in Texas. Their philosophy is simple: 'For every hundred soybeans fed to a cow, only three reach human bellies. We skip the cow.'

Challenges

Manufacturers are taking great strides forwards in terms of flavour and texture, two of the most important success criteria for consumers. But in addition to opportunities, there are also challenges and risks. One of the latter is the possibility that restrictions on the naming of certain products may put a brake on further progress. Some countries are planning to impose a ban to prevent meat substitutes from making any reference to animal-based meat (burger, sausage, steak, etc.) in their product titles. This discussion has already been talked out and agreed in Europe for the meat sector, but not for the dairy sector. Consequently, it is not yet possible to refer to a soya drink as 'soy milk'. Price is another difficult issue: meat substitutes are not much more expensive than premium meat, but they are a lot more expensive than the cheaper kinds of fast food.

Then there is also the question of whether or not meat substitutes are as healthy as they claim. The newest generation of veggie burgers certainly contain less cholesterol than lean meat, but they have the same amount of fat, less protein and more salt. They also have a high level of unsaturated fats, due to the use of coconut oil. Some plant-based burgers likewise contain relatively high concentrations of carbohydrates and some make use of additives, like flavour enhancers. The Impossible Burger even contains a completely new ingredient: soy heme, a substance that also occurs in the haemoglobin in human blood and is extracted from soy by fermentation. This is the substance that makes this special veggie burger taste most closely like meat ... although some consumers remain suspicious of this new technological wizardry. At the time of writing, this burger has still not been cleared for sale in Europe.

Even so, in general there is no sign that the rapid rise in the sale of plant-based meat is likely to come to a halt anytime soon. The products are becoming more sustainable, healthier and tastier, and are finding their way into an increasing number of consumers' lives via supermarkets and restaurants. But how far can this process go? Are we nearing the end of this evolution or are there still several more stops along the way? The ultimate dream of many of the substitute pioneers is that one day they will all be able to cultivate meat, poultry and fish 'in vitro' in bioreactors. In other words, meat from the laboratory or even the factory. This, they say, would allow us to eat 'real' meat without the need for a single animal to be slaughtered and with a significantly reduced ecological footprint. We will look at this again in the next chapter.

THE REARGUARD ACTIONS OF THE MEAT AND DAIRY SECTORS

Of course, the increasing interest in animal-free meat and dairy products has not escaped the attention of the traditional agro-food sector. Farming organisations see that meat and milk are under fire in public opinion – and not always for valid reasons and with valid arguments – while hip oat drinks and anything but 'natural' imitation hamburgers are praised to the heavens. This not only causes irritation, but has also provoked a serious counteroffensive. As a result, lobby organisations are increasingly attempting to block the forward march of plant-based proteins. How? Amongst other things, by putting policy-makers under pressure to impose limitations on their new competitors.

For example, in 2020 the European Parliament was required to vote on far-reaching proposals that would prohibit meat substitutes from making any reference to meat in their names. According to the meat lobby, terms like steak, sausage, cutlet and burger should be reserved exclusively for 'real' meat. They claimed that names like veggie burger or soy sausage would lead to confusion amongst consumers, an argument that was dismissed by the supporters of plant-based food as ridiculous. The majority of MEPs in the parliament agreed with them: the legislation was thrown out.

Even so, the victory of the plant-based sector was not complete. The parliament confirmed the existing prohibition that prevents dairy substitutes from making any reference to dairy-based products in their names. In other words, 'soy cream' and 'plant-based yoghurt' continue to be banned. It is not even possible to refer to your product as 'an alternative for butter'. To make matters worse, a new amendment – amendment 171 – suggested the imposition of even more stringent linguistic limitations. This would mean that terms like 'creamy' could also no longer be used to describe, say, the texture of a plant-based product. There was even a proposal that these products should be denied the use of typical dairy packaging, like milk cartons ... Eventually, the European Parliament withdrew amendment 171 – and the plant-based sector breathed a huge sigh of relief.

Proteins from the laboratory

In essence, the criticism of meat production can be easily be summarised. The argument, in its most simplified form, goes like this. We feed plants like soya, maize and barley to cows, which turn that feed into meat, which we then feed to people. This is a wasteful process, in which not only time, but also valuable space, energy and water are lost. And that is before we even mention the negative effects in terms of greenhouse gas emissions. For each 150-gram steak on your plate, the cow eats over a kilogram of food and some 2,500 litres of water are consumed. Conclusion? Just give people the plants to eat instead.

It sounds plausible. But what if you could manufacture animal proteins in a factory? What if the natural growth process could be replaced by a cultivation or fermentation process? 'In vitro' meat would reduce the emission of greenhouse gases by 80% to 90%, use 45% less energy, 90% less water and 99% less land. And not a single animal would need to die. Sounds promising? Perhaps. What's more, the idea of laboratory meat or 'clean' meat is no longer as fantastical as it was just a decade ago.

On 6 August 2013, journalists from around the world looked on with interest and bemusement as the Dutch professor Mark Post unveiled the world's first-ever laboratory hamburger in London. Not that there was any ham in it. And perhaps the seasoning was a bit on the mild side, said the three people who were allowed to sample it. But it did unquestionably taste of meat. There was, however, one major problem. It had cost some 250,000 euros to make the burger. Fortunately, billionaire Sergey Brin, co-founder of Google, was prepared to dig into his very deep pockets.

CULTIVATING MEAT – HOW DOES IT WORK?

The main elements of the production process for cultivated meat are largely the same as for the making of yoghurt and the brewing of beer. All three involve the culturing of cells. As far as meat is concerned, the starting point is muscle cells removed under anaesthetic from live cattle, pigs, chickens or fish.

The cells are then cultured (grown) in bioreactors. Just like in the body of the animal, these cells are fed with an oxygen-rich medium (the serum) that contains basic nutrients such as amino acids, glucose, vitamins and minerals, supplemented with proteins and other growth factors. After a time, the cells start to reproduce themselves.

The serum represents a serious problem in this process. At the moment, it is often made on a basis of calf's blood. In the long term, this is not an option for a product that wishes to project itself in the marketplace as being 'animal friendly'. Plant-based alternatives do exist, but they are expensive. For the time being, too expensive.

Through changes in the composition of the serum, the cells differentiate themselves into muscle tissue, fat tissue and connective tissue, following which they are harvested to make end products. One muscle cell makes it possible to culture a thousand billion new cells, which the manufacturers allow to grow together until they form a hamburger or – much more difficult – a piece of steak. The entire process takes between two and eight weeks, depending on the type of meat that is being cultivated.

Initially, the culturing or cultivating of meat in this manner was done on a very small scale in petri dishes. The intention is that it will eventually be carried out on an industrial scale in huge bioreactors. This will be the only way to efficiently satisfy the consumer market.

This short video shows the production process of lab-meat at the pioneering Mosa Meat company:

https://youtu.be/kG4EO-P93Dk

The race to the marketplace

Post is nothing if not ambitious. And optimistic. He founded the Mosa Meat company and predicted that within a few years the price of one of his meat substitute burgers would drop to 11 euros per burger. He now predicts that by 2024 there will hardly be any difference between the price of cultivated meat and supermarket meat. The professor/entrepreneur is realistic enough to think that the market share of plant-based meat substitutes will remain limited, because most people will simply prefer to keep on eating real meat. 'We are going to cultivate meat in tanks that are half the size of an Olympic swimming pool. These bioreactors will allow us to produce 10,000 kilograms of meat each year,' he said in 2016 (Steel, 2016).

In 2018, Mosa Meat raised 7.5 million euros for the construction of a pilot factory that was intended to bring an affordable product to market within three years. The money was provided by M Ventures (the investment fund of the Merck pharmaceutical company) and the Swiss food processing Bell Food Group. In 2021, a new round of funding was launched, this time raising 10 million euros for the upscaling of the pilot factory in Maastricht. New investors included the Nutreco animal feed group and Jitse Groen, the top man of Just Eat Takeaway.

Anno 2021, the lab-burgers have still not reached the shelves of Delhaize or Albert Heijn. Even so, it seems increasingly likely that cultivated meat – but also cultivated fish, shrimps, chicken, milk and leather – will soon become a viable business. There is currently a huge race in progress: the race to see who will be the first to get an affordable product into the supermarkets and/or restaurants. Until recently, Mosa Meat had a small lead, but numerous competitors are now breathing down its neck, each of whom has also succeeded in securing the necessary funding for the expensive research and product development that is involved.

Memphis Meats is an American company that produced the world's first-ever cultivated meatball at the start of 2016. A year later, it followed up this success with the first chicken nuggets and the first duck meat, both cultured using animal stem cells. The company sees this as a huge step forward, particularly in view of the popularity of poultry meat worldwide. Memphis Meat can count on the support of rich celebrity investors like Richard Branson and Bill Gates, but the meat giant Tyson Foods also has its finger very firmly in the pie. This huge financial backing alone probably puts Memphis Meats in pole position to be the first to take a commercial product to the market, although the company itself has warned against the danger of rushing things. They are aware that you only get one chance to make a good first impression.

The method of production developed in Israel by Aleph Farms differs fundamentally from the approach adopted by Mosa Meat and Memphis Meats. The start-up is not interested in the production of a beef mince substitute but is focusing instead on cultivated steak. Its big breakthrough came in February 2021, when it succeeded in making a piece of entrecote using 3D-technology. According to reports, the steak was juicy, flavoursome and had the right texture. Aleph Farms has the capability to allow different kinds of cells to grow together to create complex forms and hopes to see a limited market launch of some of its products, with different kinds of meat, in 2024.

Peace of Meat, a Belgian developer of cultured meat products, was acquired by the Israeli Meat-Tech 3D for 15 million euros. This young start-up, which was founded as recently as 2019, specialises in producing animal fat without animal suffering. Meat-Tech 3D is a specialist in the 3D printing of plant-based meat substitutes, but hopes to supplement this with Peace of Meat's cultured fat. The aim is to launch a combined product in 2022.

In California, BlueNalu focuses on the production of substitutes for fish and seafood. In particular, it wants to find alternatives for overfished varieties of marine life and for varieties that are difficult to cultivate in the traditional manner or have become contaminated (for example, with mercury). BlueNalu is currently building a pilot factory in San Diego. Close behind is Finless Foods, which also wishes to cultivate fish and particularly blue fin tuna, a popular but endangered species.

According to its own publicity, the Berlin start-up Bluu Biosciences is the first company in Europe that specialises in the development and production of fish cultivated in bioreactors on the basis of fish cells. Fish is a popular source of protein throughout the world, but overfishing and pollution pose a serious threat to long-term supplies and aquaculture is unable to provide a sustainable solution, say Bluu Biosciences. Laboratory fish also has a number of advantages in comparison with cultivated meat: its structure is less complex and it can also be cultured at lower temperatures. Bluu wants to bring its first products to market in 2022 and is currently working on both fish balls and fish paste. These products, it is claimed, will contain more healthy omega-3 fatty acids than wild fish or cultivated fish. In collaboration with colleagues like Mosa Meat, Aleph Farms and Meatable, Bluu has founded Cellular Agriculture Europe, with the aim of facilitating the approval of laboratory fish and meat throughout Europe. The companies involved realise that this will require maximum transparency, if they wish to win the confidence of both the regulators and consumers.

VEGGIE BURGERS ARE STILL MORE ECOLOGICAL

What is the environmental impact of culturing one kilogram of meat in a commercial production scenario? That was the central question in a study carried out by CE Delft on behalf of GAIA and The Good Food Institute (GFI). The study made a comparison between cultivated meat and slaughterhouse meat. On condition that the cultivated meat is produced using renewable energy – bioreactors use a lot of electricity – the environmental impact would be reduced by 93% in comparison with conventional beef, 53% in comparison with pork and 29% in comparison with chicken. Even so, cultivated meat still has a bigger ecological footprint than plant-based protein alternatives.

'If you want to do what is best for the environment, you need to eat as many plant-based foods as possible. If you still want to eat meat, then it is better to opt for cultivated meat made with green energy. That is good both for the animals and for the environment,' says Hermes Sanctorum, GAIA's cultured meat consultant. According to him, it is illusory to think that everyone will one day eat exclusively plant-based food. 'The consumption of meat continues to increase worldwide. Even in Europe and Belgium, where people now eat less meat, more meat is actually being produced for export. If we are really concerned about animals, we need to obtain our meat differently. Just arguing in favour of more veganism will get us nowhere.'

A taste for more?

What is the state of the market today? Companies have made considerable progress already, creating the impression that a commercial product cannot be far away. In fact, the first restaurant that works exclusively with chicken from a laboratory is already in operation. It is the brainchild of the Israeli producer SuperMeat and opened its doors in Tel Aviv at the end of 2020. Called simply 'The Chicken', it is the world's first test kitchen for chicken meat cultivated from chicken cells.

From their tables, the guests have a fine view of the production facility and also have the added bonus that they don't need to pay for their meals in cash, but with feedback about the food. The most popular item on the menu is the cultured chicken burger, which, according to the test panels, is indistinguishable from a conventional chicken burger. This kind of testing is essential for the producers of meat substitutes: if they want to be successful, they need to convince the taste buds of their consumers.

SuperMeat produces chicken meat using stem cells, which in theory have the inherent capacity to reproduce themselves endlessly. The technology was first developed in the pharmaceutical industry, but has been upscaled to allow the large volume production of meat. The process is much faster and more efficient than breeding chickens, says CEO Ido Savir. 'As soon as the desired animal mass has been reached, half of the meat can be harvested every day. It is like having a farm with a thousand chickens, five hundred of which you can cull daily, in a never-ending cycle.' The company hopes to start supplying restaurants in 2022 and plans to open a factory on a commercial scale in 2025. When this happens, the price of cultured chicken is expected to be similar to that of real chicken.

You can view a short report about The Chicken test restaurant here:.

https://youtu.be/baF22S1Yg2c

Four major challenges

It is clear that the sector is slowly moving to maturity. An ecosystem is gradually being developed around the world of cultivated meat. More and more companies are emerging to make the necessary bioreactors or the serums that are needed for cell culturing. In the early days, the pioneers had no option but to discover everything for themselves, which slowed down the development process. Nowadays, the speed of that process is accelerating steadily. Even so, a number of major challenges remain. Here are the four most important ones.

First and foremost, there is the question of production cost. It still costs thousands of euros to produce a kilogram of in-vitro meat. This is largely attributable to the price of the serum that is necessary to feed the cells, which can cost more than 100 euros per litre. Even so, it still seems possible that by 2030 in-vitro meat will be produced at a wholesale price of 5.66 dollars (4.70 euro) per kilo, according to the calculations in a study by CE Delft (Vergeer, 2021). This is very much a 'best case' scenario, which assumes that huge steps will be taken to reduce the cost of the ingredients for the culture medium and the manufacture of bioreactors.

The next issue is the scale of production. Until now, this production has taken place in test laboratories or in small pilot factories. Major investment will be necessary to upscale the production facilities to a size that is capable of producing the large volumes that the commercial market will demand.

This type of investment involves massive risks, because cultivated meat, as a so-called 'novel food', still requires formal official approval before it can be brought to market. This means that in Europe the sector has to convince the European Food Safety Authority and in America the Food and Drug Administration and the United States Department of Agriculture. Completing the necessary tests and complying with the necessary procedures can take a number of years and the associated costs are huge. Until now, there is only one country in the world that has opened its doors to cultivated meat: the city-state of Singapore, which approved its commercial sale at the end of 2020. There is, however, a very good (and self-interested) reason for this: the island needs to import 90% of its food and wants to become at least 30% self-sufficient by 2030 (see inset). The meat producer Eat Just has already concluded partnerships with local manufacturers for the culturing of chicken cells for the development of end products, which will initially be sold exclusively to restaurants. In December 2020, the '1880' restaurant was the first to offer cultivated meat on its menu. The United Arab Emirates are also showing a close interest in this kind of food production and for much the same reason as Singapore. Formal approval seems likely in the not-too-distant future.

The final and perhaps the most important obstacle is the consumer. Setting up the necessary infrastructure and satisfying all the procedures is difficult, but not impossible. But trying to convince people is an extremely delicate task. How can you create a market? First and foremost, the products need to be good. Very good. But even if they are, it will still be necessary to overcome the scepticism in some quarters about the sophisticated technology involved. Do you remember the huge level of public resistance to genetically modified crops? The concept of laboratory meat also seems equally science fiction-like to many consumers and the use of terms such as 'in-vitro' and 'cultured' does little to improve the public's perception of these products. In other words, making the breakthrough for lab-meat is not only a challenging task for the bioengineers, but also for their marketing colleagues. Winning consumer trust is essential. Transparency is a must. One of the key questions is whether vegetarians and vegans can be persuaded to eat the new products, since they no longer involve suffering to animals. The initial reactions from these groups have been mixed.

What kind of time frame are we talking about? Most companies claim that they will be ready to go to market with a commercially viable product by 2025. At least as far as the production-technology side of the equation is concerned. But everyone is aware of the risk of being overhasty: the launch cannot afford to be a failure. One incident can be enough to saddle the entire sector with a bad reputation that it will take years and billions to repair. No-one is willing to take that chance.

CULTURED MEAT IN SPACE?

Like vertical farming, it seems that in-vitro meat might also have a future in space. For astronauts, fresh food is a rare luxury. If they could culture stem-cell meat in their space station – or who knows, later in their moon or Mars colony – this would add some welcome variety to their diet. And we now know that this is no longer science fiction: in 2019, the Israeli start-up Aleph Farms cultivated beef cells on a small scale in the International Space Station.

Milk without cows and eggs without chickens?

Consumers who wish to enjoy milk, yoghurt or cheese that involves no animal suffering currently need to turn to soya or oat drinks and cheese substitutes made from cashew nuts. These products are getting better, but for many people the difference with the real thing is still too great. But what if milk could be cultured in the same way that some start-ups are culturing meat? The good news is that it can! What you can do with meat cells, you can also do with milk cells. In fact, it is even easier. And this is because there is an essential difference: milk proteins can be cultured through the fermentation of yeasts: you don't even need a cow!

One of the pioneers in this field was the Silicon Valley start-up Muufri (pronounced as 'moo-free'), which has since been renamed as Perfect Day. The company focuses its efforts on an ingredient that gives so many dairy products their unique flavour: whey. They were the first to market a non-animal whey, produced via a fermentation process involving yeast and bacteria that resulted in a protein with a real dairy taste. Perfect Day now supplies this core ingredient to other manufacturers of dairy alternatives.

One of its customers is Brave Robot, a producer of dairy-free ice-cream that launched its products in 5,000 American supermarkets in the spring of 2021. These products taste like real ice-cream, but not a single cow was involved in their manufacture. Once again, the proteins were obtained by a micro-biological fermentation process. New Culture, another new start-up, has opted for a different approach, concentrating on cheese as its end product and using laboratory-cultured casein to create it. The first tests are said to have resulted in a very tasty and highly elastic lab-mozzarella.

One of the main (and important) differences with in-vitro meat is that milk proteins are not cultured by using 'real' milk, but on the basis of fermented yeasts. Consequently, Perfect Day regards its products as perfectly vegan, even though its proteins are exactly the same as milk proteins. This means that these products can be brought to market without the need for exhaustive and expensive approval procedures. In fact, many of their methods are not new at all. Take rennet, for example. This used to be extracted from the stomach of calves, but since the 1990s it has been produced using bacteria. Just like insulin, which used to be extracted from pigs but is now made in a laboratory (Carrington, 2018).

Moreover, what you can do with milk, you can also do with eggs, and in precisely the same way: through a fermentation process that provides proteins that are indistinguishable from real egg proteins, but with not a chicken in sight. The process is actually relatively simple, at least as far as the proteins are concerned. There are 'only' twelve different proteins involved, and each of them can be cultured with the help of bacteria. One of the companies interested in this field is mega-brewer AB InBev, which through its innovation division ZX Ventures has invested in a number of companies involved in perfecting precision fermentation techniques for the creation of animal-free proteins. One of these companies is Clara Foods, which specialises in vegan proteins. The link with AB InBev is not hard to find: brewing beer is also a matter of fermentation.

But if making egg proteins is reasonably straightforward, making the egg yolk is another matter. Yet despite the complexities, there are already a number of good imitations on the market. For example, the Indian company Evo Foods recently launched a liquid egg alternative, with a flavour and texture that are very similar to those of eggs laid by chickens. They claim that their product contains more proteins than traditional eggs and is made from mung beans, chickpeas and peas. The start-up from Mumbai has deliberately opted for ingredients that are readily cultivated in India, so that they do not need to rely on imports. And with a domestic market of 1.4 billion people, they probably won't need to export their end product either! Evo is the first vegan egg substitute in Asia, just beating OnlyEg, which will make its market debut in Singapore in 2022.

There is one final question that needs to be addressed: laboratory food might be more sustainable than traditional products from the agro-industry, but is it also as healthy? Because lab-food makes use of the same proteins and same cell structures, the answer to this question is 'yes': there is hardly any difference. Lab-meat and lab-milk also have almost the same nutritional value as their real counterparts. In fact, it may be possible to further optimise and adjust the nutritional value of lab-food, opening up the future prospect of highly personalised foodstuffs. More about this later. For now, suffice it to say that cultivated meat and cultured milk also have the added benefit of making no use of antibiotics, although they cannot be regarded as diet food: a vegan ice-cream also contains plenty of fat and sugar.

WHY SINGAPORE IS PLACING ITS FAITH IN URBAN AGRICULTURE AND CULTIVATED MEAT

One of the frontrunners in the field of urban agriculture – and not only in that field – is Singapore. The densely populated city state produces just 10% of its food needs, making it heavily dependent on imports and vulnerable to the disruption of its food supply chain, as became painfully evident during the corona pandemic. For this reason, the government wishes to ramp up its own local production of food, with a target of 30% self-sufficiency by 2030. Urban agriculture is one of their spearheads. Today, just 1% of Singapore's surface area of 724 square kilometres is used for agricultural purposes.

This needs to increase – and fast. Consequently, the state authorities have invited tenders for the creation of city farms on industrial sites and on the roofs of public car parks (Aravindan, 2020), as well as financially supporting the setting up of covered farms. In collaboration with Panasonic, the agro-start-up IFFI (Indoor Farm Factory Innovation) is planning to set up this kind of farm for the cultivation of 3,500 square metres of vegetables at twelve different levels, resulting in a daily harvest of 800 to 1,000 kilograms. It is also intended that this farm should become a training and research centre to encourage others to do likewise, while a nearby 'farm-to-table' café must excite public interest (Chong, 2020).

Given this policy, it is no coincidence that Singapore was the first country in the world to give formal approval for the commercial production of cultivated meat. According to the city state's food agency, the cultured chicken produced in San Francisco by the well-established start-up Eat Just complies with all the country's safety norms and can therefore be used for the manufacture of chicken nuggets. Eat Just has already concluded partnerships with local factories to culture the necessary chicken cells and to develop the end product, which will be sold in the first instance to restaurants. The company also hopes to obtain a permit for the sale of its chicken fillets.

FERMENTATION, A TECHNOLOGY FOR THE FUTURE

In the search for healthier and more sustainable methods of food production, fermentation is coming increasingly to the fore. Consumers are familiar with the principles of fermentation from sauerkraut or its hip South Korean variant kimchi and the production of the meat substitute Quorn is also the result of a fermentation process. However, there are plenty of other possible applications in the food industry, some of which are already being used.

Fermentation is a technique that makes use of algae or fungi to produce nutritional additives and/or ingredients: natural colourants and flavourings, fats, proteins, etc. Fermentation can also improve the texture and mouth sensation of plant-based alternatives for milk and cheese. The secret heme ingredient that makes the Impossible Burger bleed like real meat is likewise a fermentation product.

Additives produced by fermentation are more natural and healthier than additives based on petroleum derivatives, which are still often used in sweets, mints, cakes and snacks. The production process requires little or no additional use of land, water and energy. This is what potentially makes it a real game changer. Many food experts believe that biofermentation could be on the eve of a major breakthrough.

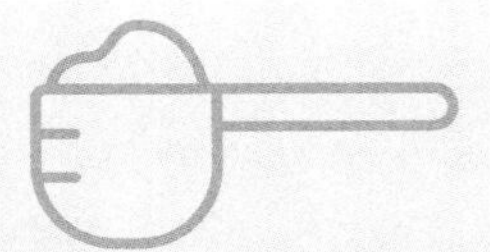

WILL WE ALL BE EATING HOT AIR IN 15 YEARS' TIME?

Microbial protein is another potential solution that can help us to make the necessary protein transition. Bacteria can be used to culture proteins in bioreactors. The main advantages are the fact that the cultivation process is fast, does not require the use of scarce and expensive land, and can recycle waste water from the food production industry. Likewise, it can also make use of biogas from the natural fermentation of food waste or, alternatively, CO_2. Researchers have discovered protein-producing microbes that contain 70% more protein content than soya beans, whilst at the same time absorbing both CO_2 and atmospheric nitrogen. In other words, a circular system.

Microbial proteins can be used, for example, as fish feed in aquaculture (its cellular structure is similar to that of fishmeal). The cell walls of bacteria also contain substances that can make animals more resistant to disease. Tests with purple bacteria have revealed that shrimps grow faster with microbial protein than with standard fish food. Moreover, these proteins are also suitable for human consumption; for example, as an ingredient in meat substitutes.

But perhaps the most imaginative and thought-provoking application is currently being developed by the Finnish start-up Solar Foods, who are experimenting with a new kind of nutrient that almost quite literally is plucked from the air. This new protein, based on hydrogen and soil bacteria, is ten times more efficient than soya, with which the company hopes it can compete from 2025 onwards. By mixing the bacteria with hydrogen, a kind of previously unseen meal is created that might well become a staple for the food of the future. This idea was originally investigated in the 1960s as a possible source of space food, but it has only recently been put into practice for the first time in the laboratory of Solar Foods in Helsinki (Harrabin, 2020).

Water is broken down into its constituent hydrogen particles using electricity and the end result is a yellowish protein meal that has neither smell nor flavour. But that is precisely the intention, since it means that the protein can be used in almost every kind of foodstuff. The developers are already looking at possible applications in bread and pasta, but it can also potentially replace the use of palm oil in sauces, biscuits and so much more.

It is also predicted that the meal powder, known as Solein, will play a major role in the meat industry. Solar Foods expects its product to match the cost price of soya by 2025, which makes it an interesting alternative for animal feed. But even better for the environment and our over-exploited planet is the fact that the new protein can be used as a basis for cultured meat and fish.

Because the process requires almost no space and no natural resources, the Finnish scientists claim (not without justification) that their system is hyper-efficient and super-sustainable. This system is, however, dependent on the ready availability and low price of sustainable electricity, since electricity is a crucial part of the process. During a first investment round, Solar

Foods was able to raise 5.5 million euros, which will be used for the product's further development and commercialisation.

Solar Foods' CEO Pasi Vainikka talks about this revolutionary new protein in the following video:

https://youtu.be/egjWeGFdrso

Insects: the food ingredients of the future?

In 2014, they were briefly a hype in Benelux supermarkets: burgers, nuggets and spreads on the basis of insects. You could find them in Delhaize, Albert Heijn and Carrefour. Natural food specialist Damhert even created their own brand: Insecta. At smart receptions, the guests were offered paprika-flavoured deep-fried locusts as a finger food with their cava. The food pundits were queuing up to tell us that mealworms and crickets were destined to be the food of the future. After all, they are a high-value source of proteins, energy, fats and minerals, which makes them an excellent sustainable alternative to meat.

After just two years, these products disappeared from the shelves. Sales were disappointing. 'We had some success in the beginning. People were curious to try them. We also tried to give them a prominent position in the market; for example, through regular tastings in our supermarkets. But the vast majority of people just weren't interested. The insects aren't selling and so we are withdrawing them from our range for the time being,' explained a Delhaize spokesperson in 2016. Since then, things have been pretty quiet around this new food category that never quite lived up to its expectations. Are they gone for good? Time will tell.

'The problem is between people's ears'

The biggest problem with insects as food is between the ears of consumers. It is first and foremost a mental thing. That, at least, was the conclusion of a study carried out under the leadership of Professor Xavier Gellynck of the Faculty of Bio-engineering Sciences at the University of Ghent. PhD student Joachim Schouteten investigated what a test public of almost 100 young people thought about eating insect burgers. To help him, he invented what he called an Emo-Sensory Wheel, an interactive tool on which the test subjects could indicate their emotions in response to different kinds of food, but could also register sensory properties like after-taste, juiciness and colour.

'The tests showed that young people who ate the insect burgers without any prior knowledge found them much less tasty than a conventional meat hamburger, but had no problem to eat them,' comments Xavier Gellynck. 'However, as part of the test they were also asked to say what they felt about the prospect of eating an insect burger. Fear and distrust were their most prominent emotions.' The majority of the participants also found the insect burgers to be dry in comparison with their meat equivalents (Gellynck, 2006).

Surprisingly enough, the insect burgers actually scored better when the young people knew in advance what they were eating; namely, a burger consisting of 31% of mealworms. In other words, the level of distrust was reduced when people were better informed. Most of them also said that they regarded insect burgers as being healthier than meat burgers. These are important conclusions for the possible future marketing of foodstuffs based on insects.

WHAT IS SO GOOD ABOUT INSECTS?

To start with, they are good for your health, being rich in fats, proteins, vitamins, fibres and minerals. They also contain essential amino acids and omega-3 fatty acids, while some species also have high concentrations of iron.

The cultivation and consumption of insects is more sustainable than livestock farming. Insects convert their food into edible body mass much more efficiently than cattle. Two kilograms of feed yields roughly one kilogram of insect 'meat', 80% of which is edible. In comparison, a cow needs eight kilograms of feed to produce one kilogram of meat. Moreover, we can only eat 40% of a cow. In addition, the cow grows much more slowly, and requires much more land and water than insect farming. Last but not least, insects live on what is otherwise waste food and they emit almost no greenhouse gases. What more do you need?

Almost forgot, there is also this: insect farming provides much needed employment in many poorer regions and without the need for investment in sophisticated installations.

Small-scale insect farms

Viewed in global terms, eating insects is nothing out of the ordinary. According to the United Nations Food and Agriculture Organisation, more than two billion people worldwide regularly have insects as part of their diet, especially in Africa, Asia and South America (Van Huis, 2013). There are more than 1,900 edible species, of which the most commonly consumed are beetles, caterpillars, bees and wasps. Even here in the West, we eat insects every day, although without knowing it: the remains of insects are inevitably processed in, for example, beer, fruit conserves and dried spices.

Most insects are harvested in the wild, although some species, such as honey bees and silkworms, have been domesticated for thousands of years. Other insects are grown so that they can be used in horticulture, either as pollinators or as the natural predators of more harmful species. However, the idea of growing insects as a source of food is much more recent. Nowadays, cricket farms are not uncommon in countries like Vietnam or Thailand. Mealworms and locusts are also grown in a similar way. These farms are usually small-scale family enterprises; production on an industrial scale is still comparatively rare. Most of these farmed insects are intended for direct human consumption, although they can also be processed into meal or paste for inclusion in more complex dishes. Extracting component substances from the insects, such as proteins, minerals and vitamins, is technically possible but very expensive.

Clever concealment

Most Europeans regard entomophagy (a difficult word that means 'eating insects') as a primitive and generally revolting eating habit. Of course, that is one of the key things about eating habits: they are culturally determined. Some cultures eat dogs, although this is seen as incomprehensible by almost every European. Nearer to home, some countries and regions – France and Flanders, for example – like to eat horsemeat, while others view this as little short of disgusting. In other words, it is all a matter of perception. And the Western perception of insects is that they are nasty, creepy-crawly things that they would prefer not to see on their plates. Obviously, this makes the breakthrough of insects as food in Europe and North America something of a challenge. That being said, those who have had the courage to try them often find them tasty. René Redzepi, the chef at the world famous Noma restaurant in Copenhagen, has used them regularly.

Perhaps insects can win the favour of Western consumers if they are cleverly concealed in more attractive products. One interesting exponent of this idea is Kriket, a Brussels start-up that markets hip and smartly packaged snack bars based on crickets, seeds and nuts. There is also a cricket granola on their menu. The company wants to free insects from their niche status and make them a mainstream source of protein. The crickets are grown at a sustainable urban farm in the city, where they are fed on waste food. Kriket is already selling its bars in Belgium, Germany, the Netherlands, the Czech Republic and the United Kingdom, and in 2020 the Colruyt Group acquired a minority interest in the company. The retailer not only provides much needed financial resources that Kriket will use to develop new products, but also brings important knowledge and experience of the consumer market. 'With the support of Colruyt, I believe that by 2030 we can make crickets as much a normal part of people's diet as shrimps or scampis,' says founder Michiel Van Meervenne.

The Brussels start-up is the first company in Europe to have been awarded a biocertificate for a product based on insects. In theory, however, its insects are not yet approved for human consumption: an EU guideline states clearly that all new foodstuffs must first be tested by the European Food Safety Authority. Some countries, including Belgium and the Netherlands, have argued that this does not apply to foodstuffs that are clearly of animal origin, but in 2018 new EU legislation decided otherwise, although a transition period was agreed. Since then, opinion in Europe seems to have changed back in favour of insects. At the start of 2021, the EFSA gave the green light for the human consumption of yellow mealworms, the first insects to be 'honoured' with this distinction. A number of other insect species look set to follow in the near future. This can give a serious boost to the sector as a whole. But a word of warning: people who are allergic to crustaceans and shellfish are also advised to steer clear of insects. You may not think it to look at them, but they are distantly related to each other ...

Food for your four-footed friends

There may be greater potential for the use of insects in animal foodstuffs, where they could form a more sustainable alternative for fishmeal and soya. Since 2017, Europe has approved the use of insect proteins for the cultivation of fish. This led to a rapid growth in production. According to the International Platform of Insects for Food and Feed, the annual production of insect protein is likely to rise to 3 million tons by 2030 (Ploegmakers, 2020). An interesting case in this respect is Entocycle, a London start-up that manufactures sustainable insect protein for animal feed by harvesting black soldier flies that are grown on a waste-food diet of fruit, vegetables and coffee grounds. And all in the very heart of one of the world's biggest cities.

Another trendsetting player in the sector is Protix, a company that was awarded the annual Dutch Innovation Prize in 2020. The company has developed a circular concept in which food waste is used to feed insects, which in turn are then used to serve as a basis for foodstuffs for cattle and fish. Protix argues that this is a huge improvement on the less sustainable animal feed options that are currently in use, based primarily on soya, fishmeal and palm oil. 'Insects help to counteract over-fishing and the cutting down of forests for soya cultivation. Our insect-growing process is fully automated. Smart algorithms and robotics are used to ensure that this process can be monitored and controlled at the level of both the larva and the pupa.'

The French company Ynsect, which claims to be a world leader in insect farming, raised 189 million euros to build the largest insect factory in the world at Amiens. The company grows mealworms to serve as a source of protein for cattle feed, pet food and fertilisers. The new factory is scheduled to open in 2022 and aims to produce 100,000 tons of animal feed each year. Director Antoine Hubert does not expect that insects will quickly find their way onto the plates of human consumers, unless for specific applications such as sports food. Insects are, however, an excellent alternative and sustainable source of protein for animal foodstuffs, which also benefit the health of the animals that consume them.

Using insects for cattle feed does indeed offer a number of 'medicinal' benefits. For example, it may be possible to reduce the level of antibiotic use, because chitin (a building block in the exoskeleton of insects) works as a probiotic. The black soldier fly also has anti-microbial properties and can help to combat diarrhoea in pigs. However, much further research is still needed in this field, as is a significant reduction in production costs. Another bone of contention is European legislation, which is running too far behind the rapidly evolving realities of the sector.

Be that as it may, the first pet foods based on insects are already on the market, primarily for dogs but also for cats. Once again, these products are recommended as being healthier and more sustainable alternatives. Larvae and black soldier flies are the most common ingredients: sustainably cultivated, fast growing and containing taurine, an essential amino acid that is necessary for both cats and dogs. By all accounts, they also find these new products quite tasty!

Learn more about the ambitions of world leader Ynsect in this video:

https://youtu.be/EqBIlzgBIUo

LEGISLATION IS HAMPERING THE HUMAN CONSUMPTION OF INSECTS

Before 2018, the EU imposed few limitations on the sale of insects as food. However, the Novel Foods directive, which came into force on 1 January of that year, has changed the situation dramatically – and not for the better. Foodstuffs based on insects are now subject throughout Europe to an approval procedure, although buffalo worms, mealworms, locusts and crickets are among a small number of exceptions that can already be sold.

In the US, companies can sell whole insects if they have been specifically raised for human consumption, but if they want to use a product derived from insects as an additive (for example, as a powder), they need prior approval from the Food and Drug Administration (FDA).

As far as animal foodstuffs are concerned, insects can currently only be used in the EU for fish feed, but not for cattle feed. Since July 2017, seven species of insect have been approved for use in aquaculture: black soldier fly, common house fly, yellow mealworm, lesser mealworm, house cricket, banded cricket and field cricket. However, for the time being these insects can only be fed with 'material of feed grade quality', such as material of vegetable origin, processed eggs, milk and other derivative products. Moreover, insect-based products cannot be fed to poultry or livestock, as a result of legislation that was hastily passed following the outbreak of so-called 'mad cow disease' at the end of the 1990s, prohibiting the use of animal proteins in cattle feed.

The European Commission is currently investigating the possibility of rescinding its ban on the use of insect proteins for feeding poultry, following a decision taken at the end of 2018 by the American FDA to allow the use of black soldier fly larvae in the US poultry sector.

Super-vegetables from the sea farm

In Asia, seaweed has already been on the human menu for many thousands of years, although in the Western food sector it is still very much a niche product. However, this is slowly starting to change. In particular, the increasing popularity of sushi is making people more and more accustomed to the idea of eating vegetables from the sea. In addition to nori and wakame, some European and American supermarkets are starting to stock other similar products based on algae and seaweed: seaweed salads, seaweed burgers, cheese with seaweed or dried seaweeds as a flavouring for soups. According to many food experts, the potential for further growth is massive. Sea vegetables are not only tasty – for example, they can bring a delightful touch of umami to your dishes – but are also healthy and sustainable.

Some 35 million tons of seaweed are cultivated and harvested worldwide each year. Just 300,000 tons of this is in Europe, where the industry is still very much in its infancy. However, this figure is projected to rise to eight million tons by 2030, according to a recent report published by Seaweed for Europe, a coalition of thirty or so industrial partners who aim to fully open up the European market by that date. By then, the total market value is expected to be some 9.3 billion euros, of which 30% will come from European-grown seaweed, with the rest being imported (Vincent, 2020). This is likely to result in the creation of 85,000 new jobs. In Europe, there is probably most potential for the use of seaweed as a supplement for animal feed or as a biostimulant for sustainable fertilisation. However, human consumption is also growing. Worldwide this now accounts for 77% of all the seaweed harvested and the market continues to grow strongly, thanks to renewed interest in the importance of healthier and more sustainable eating patterns.

Super-sustainable superfood

The beneficial environmental impact of seaweed cultivation is also significant. Seaweed takes CO_2 out of the sea and also absorbs nitrogen and phosphorous. In this sense, seaweed farming is regenerative farming. Its production requires no use of scarce land, but makes use of underwater space that is freely available near the coast, out at sea or in combination with wind turbine parks and aquaculture. It requires no fresh water, no fertilisers and no pesticides. In contrast to land farming, it does not threaten biodiversity but actually enhances it. Seaweed cultivation is therefore wholly in keeping with the sustainable development objectives of the United Nations and the European Green Deal.

Recent years have seen numerous initiatives that seek to take local seaweed farming to a higher level. In Flanders, for example, the Seaconomy consortium was set up in 2018, with the Colruyt Group as one of its partners. According to the initiators, seaweed can play a major role in local food supply and in improving the sustainability of the food supply chain. In the Netherlands, the Seaweed Centre is exploring similar possibilities, especially in the waters of the Oosterschelde, where a number of seaweed farms are already active.

It seems that the North Sea is particularly suitable for the cultivation of different varieties of seaweed with an economic potential, according to a recent Belgian study investigating the economic feasibility of this so-called 'superfood'. The most promising species are Atlantic wakame, sugar wrack, bladder wrack, serrated wrack, sea lettuce, dulse, nori and Irish moss. The study concludes that an ambitious innovation project is necessary to make possible the use of seaweed on a large scale as a natural resource for the future. In the short term (2025), seaweed can be grown on a local basis in farms off the Belgian North Sea coast for use as additives for food products and animal feedstuffs. The harvesting of seaweed from these Belgian waters should be more than sufficient to meet local demand from both these sectors.

The Colruyt Group is already convinced: the retail group is planning to set up a 4.5 kilometre-long sea farm off the Belgian coast near Nieuwpoort. In a first phase (and in addition to 2,000 tons of mussels and 100 tons of oysters), it intends to harvest 100 tons of seaweed; more specifically sugar wrack, a seaweed that does not grow in the wild but is a delicious addition to various dishes, such as salads. The added bonus? This seaweed is not expensive.

HOW HEALTHY ARE ALGAE AND SEAWEED?

This varies from species to species, but in general it is fair to say that algae and seaweed are packed full – sometimes even a bit too full – with health-giving nutrients.

To start with, seaweed contains numerous vitamins (including B1, K and folic acid) and minerals. It is a source of calcium, phosphorus, magnesium, sodium, potassium, iron and iodine. In addition, seaweed and algae both contain various types of protein and essential amino acids. This makes it possible for them to serve to a limited extent as a substitute for meat. These 'sea vegetables' are likewise rich in fibre and have high concentrations of the healthy fatty acids EPA and DHA, which are also found in fish.

One particularly important component is fucoidan, a substance not found in land plants and sometimes suggested as one of the reasons for the exceptional longevity of the Japanese, who have been major consumers of seaweed for decades. Fucoidan also supports the immune system and cardio-vascular functions.

But watch out: too much seaweed is not good for you either. In particular, you need to be careful with iodine. Likewise, an overconsumption of vitamin K is not a good idea. Some types of seaweed also have a high potassium content, which can lead to problems for people with kidney complaints. Others have high concentrations of heavy metals. As a result, picking seaweed in the wild is not recommended!

Success stories

So is there a market for this kind of thing? There is certainly an increase in interest. You can find a limited offer of seaweed in health food stores and in supermarkets like Delhaize and Albert Heijn. But as long as this offer is restricted to dried varieties of seaweed and a few fresh salads, the big breakthrough is unlikely to happen anytime soon. But the launch of attractive pre-prepared dishes and convenience food could soon change all that.

One of the seaweed success stories in this respect is The Dutch Weed Burger, a Dutch producer of meat substitutes on the basis of (amongst other things) beans and mushrooms, enriched with seaweed and micro-algae from the Zeewaar seaweed farm in the Oosterschelde. The company's products are already on the menu in more than 200 restaurants and on the shelves of more than 300 branches of Albert Heijn. 'We believe that plant-based cuisine is the key to real change. It makes no difference whether you look at it from the perspective of ending the hard times for the bioindustry, the sustainability of the economy, or even future global peace. The solution for all these problems is on your plate. The choice is yours,' says co-founder Mark Kulsdom. 'We are in the process of organising a revolution in the fast food industry, so that we can show the business world, the politicians, the NGOs, the farmers and the consumers that there are other options on the horizon, options that benefit nature, animals and people. We Weedheads believe that The Dutch Weed Burger can help to speed up this transition worldwide.'

Another good example of the potential of seaweed to serve as a basis for tasty and healthy food comes from Germany. Entrepreneurs Deniz Ficicioglu and Jacob von Manteuffel, in collaboration with universities, seaweed farmers and experts, are working on the development of innovative seaweed products. Their seaweed salads in jars, marketed under the brand name Nordic Oceanfruit, are now available in more than 5,000 stores throughout Germany. The seaweed is organically cultivated at different locations in Europe and their products are 100% plant-based, gluten-free and contain no refined sugars. Moreover, with their new start-up they plan to take things another step further: Betterfish is a plant-based alternative for tuna, made on a basis of seaweed and algae. It is planned to take it to market at the end of 2021.

Conclusion? Seaweed farming has shown that sustainability and economic success can go hand in hand. It is perfectly possible that in ten years' time supermarkets will be selling a wide range of tasty and healthy seaweed products for a wide public – and not just for vegetarians. Seaweed also has potential as a component for nutritional supplements, sustainable cosmetics and even life-saving medication for chronic illnesses such as heart disease and Alzheimer's. In other words, the future looks promising – and it is not far off.

Watch the story of The Dutch Weed Burger here:

https://youtu.be/O6w9OquQjAs

CAN SEAWEED HELP TO REDUCE THE METHANE EMISSIONS OF CATTLE?

Some experts predict that seaweed will become a new wonder ingredient in cattle feed. In tests, scientists have succeeded in reducing the methane emission of cows by 82% by adding a limited amount of seaweed to their diet. Just 80 grams a day was enough to achieve the desired effect, without any negative consequence for the animals' meat production.

Cows produce methane through the action of bacteria in their stomachs when digesting fibre-rich food, in a process that is similar to fermentation. A specific kind of red seaweed, *Asparagopsis taxiformis*, helps to counteract this process by suppressing an enzyme in the cow's alimentary canal, so that the production of methane is inhibited.

Methane is one of the most important (and harmful) of the greenhouse gases that are causing global warming. Agriculture is responsible for roughly 10% of methane emissions in the US, a large part of which comes from cows and their breaking of wind at both ends of their anatomy.

UNILEVER CLEANS YOUR HOME WITH SEAWEED

The potential of seaweed as a raw material goes much further than its role as a source of healthy and sustainable food for humans. It is also used in animal feed, in cosmetics, in ecological packaging and even as fuel. But cleaning with seaweed? That is the new idea of Unilever, which is developing cleaning products based on sea algae, with which it eventually hopes to create self-cleaning surfaces.

More than 80% of bacterial infections in people are caused by the formation of biofilms, a collection of micro-organisms that grow on all different kinds of surfaces, such as those you find in the kitchen and bathroom. It is possible to avoid this with seaweed, because seaweed is resistant to the formation of such films. It has a protective mechanism, which ensures that its surface always stays 'clean'. Biotechnologists have been able to develop a process based on sea algae that prevents dirt and microbes from attaching to other surfaces. As a result, for example, the black fungus that often grows between the tiles in your bathroom would soon become a thing of the past.

However, Unilever's plans for seaweed go much further than the consumer market. They want to farm out the technology under licence to the potentially much more profitable business-to-business market. A product that keeps surfaces clean has hundreds, if not thousands, of possible applications, from odourless shoes to permanently pristine banknotes. Talks are also taking place with the dental profession.

Unilever also foresees a number of possibilities with sustainability benefits. For example, if the product can help keep the hull of a ship clean, that ship will experience reduced water resistance, which in turn could lead to a 10% reduction in its fuel emissions.

A COOKBOOK FOR THE FUTURE

One of the many projects of Space 10, the research and design laboratory set up by IKEA, is a futuristic cookbook entitled *Future Food Today*. It is a collection of delicious and sustainable recipes from its test kitchen, which is intended to help us eat better, both for ourselves and for the planet. In other words, it is a tangible and tastable vision of how the sustainable kitchen of tomorrow might look. *Future Food Today* brings together technology, science and nutrition.

'We have made this book to narrow the gap between future food trends and what is currently happening in people's kitchens. *Future Food Today* aims to project a positive vision of the future of food and provide people with recipes that will allow them to move in that direction,' says the Space 10 team. 'Although we cannot predict the future, we know that the way we produce and consume food will need to change drastically, if we want that future to be better. The recipes in *Future Food Today* attempt to offer a solution to this problem by avoiding non-sustainable sources of protein, preferring to base its meals on micro-algae, insects and other ingredients that are better for the environment, but also delicious to eat.'

An example? One of the team's creations is 'Tomorrow's Meatballs': a variation on traditional meatballs, but made with insects, algae and cultured meat. Or what about the 'Dogless Hot-dog', a sustainable alternative for one of the classics from the world of fast food. Instead of meat, it contains dried and glazed baby carrots, beetroot and berry ketchup, mustard and a turmeric cream, roasted onions, a cucumber salad and a herb salad mix. The bun is made with spirulina – a micro-alga that contains more beta-carotene than carrots, more chlorophyll than wheat grass and fifty times more iron than spinach. Still not convinced? Then maybe you will prefer the 'Bug Burger': each slice has 100 grams of beetroot, 50 grams of turnip, 50 grams of potato and 50 grams of mealworms, all served on a white bun and garnished with a beetroot and blackberry ketchup, a chives spread and a hydroculture salad mix. Enjoy!

How 3D printing is changing the food industry

In the construction, automobile, design and fashion sectors, the impact of 3D printing is already making itself significantly felt. Put simply, 3D printers make three-dimensional objects, usually layer by layer, using digital files. This makes it possible for designers and manufacturers to produce complex structures and forms in a relatively short space of time and with a highly economical use of materials. It is possible to print a complete 3D house, but also a new frame for your spectacles and a pair of personalised sneakers. With a 3D printer, the consumer can become a producer. So how can this remarkable technology be applied in the food sector? In very many different ways, or so it seems.

A boost for creativity

At the start of 2020, the chocolate giant Barry Callebaut announced the opening of the world's first 3D print studio devoted to 'creating unique chocolate experiences'. This technology makes it possible to 3D print chocolate to scale. This means that from now on chefs can create their own marvellous chocolate masterpieces, but then have them reproduced quickly, easily and cheaply by printing, no matter how complex the design might be. The same procedure can be used for desserts, pralines, other sweets, warm drinks and cakes. The new service is targeted primarily at restaurants, hotels, coffee shop chains and other catering establishments. The company's investment, which was not insubstantial, illustrates their strong belief that 3D printing is on the threshold of a real breakthrough in the food sector.

Other major food producers are also showing a similar interest. PepsiCo uses 3D printing to test new shapes for its crisps and snacks. Hershey's designs new candy bars with it and Oreo uses it to personalise its biscuits. The Dutch spice specialist Verstegen sees a role for itself as a supplier of refills for food printing cartridges.

It was this same logic that persuaded pasta-maker Barilla back in 2016 to take its first steps into the world of 3D printing. They commissioned a 3D printer that is capable of producing a piece of pasta in just 30 seconds. The ultimate aim is to produce a whole plate of pasta in less than two minutes. Barilla also organises design competitions, in which the participants are challenged to create unique pasta shapes that have never previously been seen, which are capable of being reproduced on the 3D printer. Moreover, this is not just a question of visual aesthetics. This type of printer can also adjust flavour, texture and nutritional value, making it possible to amend pasta recipes to correspond with a person's specific dietary requirements or limitations, such as gluten-free spaghetti or ravioli.

A tailor-made meal

That is the promise inherent in 3D printing: it will allow us to personalise food in an efficient and scalable manner to meet people's specific dietary requirements. This type of printing makes it possible, based on your own biological information, taste preferences and health needs, to print off exactly the kind of food that is best for you. With this in mind, FrieslandCampina, in collaboration with the University of Wageningen, is exploring the possibility of printing milk proteins. In that way, the dairy producer could adapt its recipes to reflect the health situation or lifestyle choices of its customers. Amongst other things, FrieslandCampina wants to create a creamier mouth sensation for products with a low fat content.

Another possible application is the design of tasty food but with a softer structure that would be easier to eat for people who have difficulty in chewing and/or swallowing, such as the very old or patients recovering from a stroke. Until now, these people have been fed over-mixed recipes of unattractive food with little nutritional value. 3D printing could help to correct this, since the technology makes it possible to print off meal components of the right size and texture, which also have a look, taste and smell that closely match the authentic ingredients and provide the same eating experience. In other words, this means that on a wider scale it will be possible in future to compile a personalised and tasty menu for every meal that takes account of the individual nutritional needs of the consumer.

Hyper-personalised sushi

Just how far this process can be taken is already being demonstrated by Sushi Singularity, a remarkable futuristic restaurant in Tokyo. The restaurant serves 3D printed sushi that is hyper-personalised in accordance with the health and nutritional requirements of each individual guest. People who make a reservation are asked to submit biological samples, using a test kit provided by the restaurant. The biometric and DNA data obtained from these samples are used to prepare a personalised infusion of nutrients, following which the 'coded' sushi, adapted to the dietary needs of every guest, is artfully produced using 3D printers and laser technology. This extraordinary concept is a project of the Open Meals company, whose stated objective is to digitalise the entire food industry. This start-up develops cartridges for food printers, with sustainable ingredients such as seaweed, crickets and nutrients, which can be combined to take account of every health profile. Open Meals is also working on an operating system to design food digitally.

NASA bakes pizza

Science fiction? Not a bit of it! The American space agency NASA also sees potential in 3D printing as a way to add more variety to the diet of astronauts in space and, later, to colonists on the moon and Mars. A NASA spin-off is investigating how it might be possible to produce nutritional meals in the extreme conditions of outer space using 3D technology. By reducing all the component ingredients (such as carbohydrates, proteins, macro- and micro-nutrients) to powder form, their longevity can be guaranteed during the long periods associated with space travel and exploration. The users can then mix these ingredients to create a varied range of nutritional meals. Like what? Like pizza! The system first prints a layer of dough, then a layer of tomato sauce (a powder mixed with oil and water), followed by the topping of your choice, using proteins that are either animal, vegetable or dairy- based. This is something very different from the boring food out of sachets that the astronauts on the International Space Station are currently required to eat. And who knows where it might end: the Israeli company Aleph Farms is already using 3D printing for the production of cultivated meat and has recently printed off the world's first laboratory steak.

Waste becomes taste

3D printing also makes possible sustainable applications for the processing of food waste. That is the mission of Upprinting Food, a start-up founded by industrial design students at the Technical University of Eindhoven, under the motto 'Giving food waste a second taste'. Their concept reuses food that would otherwise be thrown away: old bread, but also fruit and vegetables that are no longer sufficiently attractive for sale. They turn all this waste into a purée, which is put into a 3D printer as a basis for creating new shapes and forms, which are then baked and dried to increase the storage life. The result is a series of sustainable crackers and biscuits. In the first instance, Upprinting Food hopes to sell the idea to restaurants, allowing chefs to be creative with their own waste products. However, some supermarkets are also starting to see potential in the start-up. And it is not beyond the bounds of possibility that one day we will all have a 3D printer in our own kitchens, so that families can recycle their own waste ad infinitum.

Printing in the kitchen

This is not as far-fetched as it might sound. A 3D printer for use in the kitchen already exists. It is called the 'Foodini' and its makers hope that it will one day take its place alongside the blender and the mixer in every household. This printer does not make use of standard patterns and cartridges in the way that other 3D printers do. Instead, it has five containers that you can fill for yourself with fresh ingredients: dough or sauce, sweet or savoury, etc. After that, you just print your own creations straight onto your plate or bake them off in the oven. This means that amateur cooks can experiment to their heart's content with new shapes, flavours and textures. The related smartphone app allows you to make your own designs or to use existing ones. Sadly, this new printer is not cheap, but if you have 4,000 dollars to spare you can add something truly unique to your home ...

You can see how the Foodini works here:

https://youtu.be/EO-WRFx_j20

THE 3D PRINTED VEGETABLE GARDEN

Chloé Rutzerveld calls herself a 'future food designer'. This young Dutch woman devises ground-breaking new concepts that offer a completely different way of looking at food. 'By daring to look ahead and focusing on alternatives for the future, we can get a new perspective on what we eat, why we eat it, and what we will (or will not) eat in the years to come,' she says. For one of her projects, Edible Growth, she investigated ways to create an eatable ecosystem using a 3D food printer. You can imagine it as a kind of fully edible mini-vegetable garden. Multiple layers containing a basic support structure, an edible breeding ground and various organisms are printed directly inside a tiny reusable greenhouse of carbohydrate, according to a personalised 3D file. All the consumer then has to do is place the greenhouse in a sunny spot on a window ledge. The natural process of photosynthesis does the rest: within three to five days the plants and mushrooms are fully grown. The intensity of the taste and aroma increases as the 'garden' ripens, which is also reflected in its changing appearance. The consumer can decide when to harvest and eat the garden according to their preferred degree of intensity. Short chains don't get any shorter than this! And the system is also natural, healthy, has no waste and needs no packaging.

Do you want to see how wonderful it looks?

https://youtu.be/hw321SwC6kA

Towards a healthier and more personal diet

Try to imagine that you are living in the year 2050. Every citizen has a health ID-card, a kind of digital fiche on which all your individual health characteristics are recorded: the history of all your illnesses, chronic medical conditions from which you are suffering, specific problems to which attention needs to be devoted (too much cholesterol, vitamin or calcium deficiency, etc.), your allergies (if any) and other relevant matters that have been identified on the basis of your blood group, microbiome and DNA. The fiche will also monitor your mental health and well-being: are you sleeping well, have you been down in the dumps recently, etc.? Medical research is making it increasingly clear how the different systems in our body work together and, more particularly, how the brain, the digestive system and our emotions are all connected to each other. Taken together, all this data forms your unique health identity, which, in future, will be capable of conversion into a single code. When you scan this code into the terminal at the entrance to your work canteen, a food printer will print out a highly personalised meal for you, which will contain exactly the right combination of components that you require at that moment. Exactly the right number of calories and exactly the right amount of fibre, carbohydrates, proteins, minerals, vitamins and other nutrients. Let's hope that it also tastes good!

Is this the direction in which we are travelling? It is certainly not a new idea: 'Let food be your medicine and let medicine be your food.' This famous quotation is usually attributed to the Ancient Greek physician Hippocrates, who is generally regarded as the founding father of Western medicine. There is some doubt about whether he actually said it, but that doesn't really matter. It shows that even in ancient times attention was focused – and rightly – on the importance of eating habits. Old cookbooks often contain not only recipes for meals, but also remedies for curing various illnesses, based on herbs and other ingredients. However, this popular knowledge has largely been lost following the professionalisation of medicine in recent time.

That being said, the boundaries between pharmacy and the food industry are once again starting to become vague. The importance of sickness prevention and the role that healthy eating patterns can play are once again coming to the fore. Food companies are starting to help consumers to make healthier choices. Interest in the health aspects of food has never been greater than it is today. People are ever more aware of the consequences of their food choices, sometimes if only out of necessity: in recent decades the number of food allergies and intolerances has increased alarmingly. Similarly, diabetics are not free to eat whatever they like whenever they like, while people who suffer from a nut allergy also need to be very careful about the content of their food.

Other food choices are often a matter of lifestyle. In supermarkets, the number of gluten-free products continues to rise significantly, to the extent that their market share is now much greater than can be explained by the limited number of people who are actually affected by coeliac disease. Restaurateurs will confirm this: nowadays there are very few diners who order the menu of the day without some kind of corrective comment or request. One person wants steak with pepper sauce, but the sauce must be made without cream, because of their lactose intolerance. Someone else wants the cod with broccoli, but without the mashed potato, because they are on a low-carb diet. A third guest might be a vegan, who expects a 100% plant-based alternative! And the chef and his kitchen team are expected to provide it all, and will lose customers if they don't! It speaks volumes that in the US it is now becoming trendy to list your food restrictions on your visiting card.

Recent research has shown that some foods and nutrients have the potential to strengthen the power of the human brain, improve intestinal health, regulate our moods and keep our hormones in balance. Because people in today's society are constantly seeking to improve their performance in both their private and professional lives, interest is growing in these foodstuffs and the related extracts, powders and derivatives that are believed to have health-enhancing properties.

Supplements and shakes

As a food trend, health has long been at the top of the agenda for both consumers and the food industry. Its importance has been even further increased by the corona epidemic. On the one hand, some people wanted to find solace in what have been called 'forced health' products. These are products that are actively health-enhancing, such as food supplements or probiotics that strengthen the intestinal flora and natural immunity. On the other hand, there was also unmistakeably a move towards 'soft health', which resulted in the increased consumption of fresh fruit and vegetables, legumes and organic food. Another related trend, and one that has been around much longer, is 'clean food', in which people try to avoid harmful artificial additives (preservatives, colourants, flavourings, etc.) and potential allergens.

All these phenomena are a reaction against industrialised food, which is becoming ever more intensely processed, with long lists of ingredients that few people have ever heard of, printed in microscopic letters on the packaging. The growing preference for plant-based food and the rise in the number of flexitarians and vegans are also partly driven by the same health considerations. What used to be fast food now needs to be both fast and good: quick snacks are not disappearing, but nowadays they need to be increasingly healthy and responsible. Many are even getting culinary ambitions, thanks to the influx of street food influences from all around the world.

For another group of consumers food does not necessarily need to be a sensory and/or social experience. The only thing that interests them is consuming in the most efficient manner possible the necessary nutrients that will allow them to perform in a demanding environment. Perhaps the most well-known product catering to these people is Soylent, which provides the necessary daily amounts of vitamins, minerals and macro-nutrients in a single shake. It should come as no surprise that Soylent was developed in the high-pressure world of Silicon Valley, where it offers the perfect answer to the wishes of tech-entrepreneur Elon Musk (Tesla, SpaceX) and so many others like him: 'If there was a way I didn't have to eat, so that I could work more, I simply would not eat. I wish there was a way to get nutrients without sitting down for a meal.' The company also believes that its high-tech shakes may eventually be a solution for hunger in poor regions, where fresh food is scarce – although at the present time the production costs are still far too high to make this viable.

The elephant in the kitchen

Nobody can deny that there is a problem with food and health. Hunger and food shortages still plague a significant proportion of the world's population, while in the West – much to our shame – obesity is the elephant in the kitchen. According to the World Health Organisation, the problem of obesity has tripled since 1975. More than 650,000 men, women and children, equivalent to 13% of all the people on Earth, were categorised as being obese in 2016 (we were unable to find more recent figures). Some 1.9 billion adults, equivalent to 39% of people older than 18 years of age, were overweight. Obesity is also increasing among the under-18s at an alarming rate.

At the same time, the rising incidence in the number of so-called 'civilisation diseases', such as diabetes, heart conditions and cancer, can be attributed at least in part to changed patterns of eating in the industrialised world. We eat too much and too many of the wrong things. This makes it easy to point the finger of blame at the food manufacturers, who continue to tempt us with affordable, irresistible but not very healthy and heavily over-processed goodies, most of which contain too much sugar, salt and/or fat and too little real nutrition. Public enemy No.1 is the soft drinks industry, followed closely by the snack producers and fast food chains.

However, it would be nonsense to claim that the food industry is consciously trying to undermine people's health. The food industry does what every industry tries to do: bring successful products to market. You could even say that they are just too good at their job: they produce delicious products at a reasonable price that we are simply unable to resist. What's more, you can find them everywhere – in the West, you are never far away from a local store, snack bar, supermarket or vending machine – and there is masses of persuasive marketing (more people doing their job well) to convince you that you can't live without them.

The irresistible trio

What is it that makes some industrial food so unhealthy? During the 1980s, fat – and in particular saturated fat – was seen as the main evil-doer. This resulted in a brief fad for 'light' foods: the manufacturers took all the saturated fats out of their products and replaced them with trans fats (which we now regard as dangerous!) and added sugar and salt to enhance the flavour. Even today, many products still contain so-called 'hidden' sugars; hidden because you might expect sugar in a biscuit or cake, but not in soup, sauce, pizza, crisps and cold meats.

The link between sugar, obesity, cancer and heart disease has now been irrefutably proven. Moreover, there are increasing indications that sugar – especially in combination with fat and salt – has addictive properties. Unfortunately, most successful snacks and fast food items have perfected the delicate balance between sugar, fat and salt in their recipes and also play cleverly with textures (smooth versus crunchy). The result? After eating one Big Mac, you order another one. And surely you are not going to leave that bag of crisps half eaten? Or that Kit Kat?

If people are truly powerless to resist these temptations, education and good arguments will do little to change the situation. Governments are becoming increasingly alarmed and continue to launch awareness campaigns, but they fail to impose sugar taxes or to limit advertising. But perhaps the first signs of change are now on the horizon. The British prime minister Boris Johnson, who readily admits that he is not exactly a role model in these matters, declared war on obesity in 2020 by introducing restrictions on the promotion of food products that contain too much sugar, salt and/or fat. This was a surprising move, coming as it does from a politician who has always been opposed to too much state interference in people's lives, but it underlines the seriousness of a problem that can no longer be denied. Further restrictive regulations relating, for example, to the positioning of unhealthy products – not in displays, not near the tills, etc. – could do much to change the appearance of our shops and supermarkets for the better, at least in terms of encouraging healthier food purchases.

In 2018, the UK was also one of a dozen or so countries around the world that introduced a tax on soft drinks. Opinion is still divided about the effectiveness of such taxes. However, the statistics suggest that it did help to successfully change consumption habits in the UK: in recent years the amount of sugar purchased by the average British family has fallen by 10%.

FAO SOUNDS THE ALARM

Every year, the Food and Agriculture Organisation (FAO) of the United Nations publishes a report entitled 'The State of Food Security and Nutrition in the World'. The findings of the 2020 report are not encouraging. Since 2014, the number of people suffering from malnutrition has been slowly increasing and now amounts to 690 million men, women and children, or 8.9% of the world's total population. A large part of this recent growth in food insecurity can be attributed to a rise in the number of conflicts, the effects of which are often made worse by climate-related matters. The ambition to end hunger by 2030 – one of the United Nation's Sustainable Development Goals – looks further away than ever.

At the same time, the problem of overeating is also increasing. If its prevalence continues to rise by 2.6% each year (the current rate), by 2025 obesity amongst adults will have risen by 40% in comparison with 2012, and this while the UN's objective is to halt the increase in obesity by that year. Overweight amongst children younger than five years of age has increased from 5.3% in 2012 to 5.6% – or 38.3 million children – in 2019. Poverty is one of the decisive factors. A healthy diet is 60% more expensive than an 'adequate' diet that contains sufficient nutrition and five times more expensive than a diet that only provides sufficient energy.

Pot by pot

The food industry is attempting to react to this situation. Retailers and brand manufacturers have reduced the percentage of fat, sugar and salt in their products, added more fibre, optimised nutritional values and reduced portion size. Product labels and apps provide consumers with detailed information about nutritional content and health. There is not a single food multinational in the world that does not now have 'health' as one of its key priorities, even if some of them only pay lip service to it.

Of course, the manufacturers cannot remove fat, sugar and salt from their products at a stroke: no-one would buy them. But a gradual step-by-step approach has the potential to be successful. This was illustrated in 2016 by the French Intermarché chain of supermarkets, with its 'Sucre Détox' (Sugar Detox) campaign. The retailer wanted its consumers to become more used to food with a lower sugar content and launched a number of products, including yoghurt and a chocolate dessert, in a special 'detox' version. This was a six-pack of individual pots. The first pot contained 5% less sugar than an average yoghurt, a figure that was successively increased until the sixth and final pot contained 50% less sugar. It is a clever example of 'nudging' the consumer in the direction of healthier eating, without expecting them to radically change their behaviour overnight.

There is, however, a reverse side to all of this: the so-called 'health halo'. The perception that a product is healthy can encourage some consumers to eat more of it than is intended, because 'it can't do me any harm, can it?' Classic overcompensation.

The breakthrough of the Nutri-Score

How can we make it easier for consumers to make healthy choices? Manufacturers are already obliged to list all the ingredients on their packaging, but it is by no means simple for people to correctly interpret all this information, even if they are prepared to make the effort. One possible solution to this problem is the Nutri-Score, a label that has been developed by the French Ministry of Health. The label appears on the front of food packaging and indicates by means of a letter and a colour the extent to which the product contributes towards a healthy eating pattern. Sugars, saturated fatty acids, calories and salt all have a negative impact on the score, while fruit, vegetables, fibre and protein have a positive effect. A number of European countries have now officially approved the use of this label: France, Belgium, Luxembourg, Germany and Spain. Producers are not obliged to use the label and the system remains a voluntary one.

The label has been (and still is) a source of controversy. In the beginning, most of the major food multinationals were against it, although the retailers and their private labels were quick to adopt it. In the meantime, the label has broken through in most of the food industry and large groups of scientists are now arguing vociferously that its use should be made compulsory throughout the EU. Perhaps this is a good idea, but it is important to ensure that the label is used for its proper purpose: to allow people to make a carefully considered choice of product within specific product ranges, such as yoghurts or biscuits. The Nutri-Score is not intended to help consumers decide whether deep-frozen chips are healthier than smoked salmon. It is also important to realise that the label alone is not enough to persuade people to change their behaviour: manufacturers can lower the sugar content in their biscuits, but this will not necessarily convince consumers to eat more vegetables.

Some of the southern countries in the EU are strongly opposed to making the label compulsory, because they believe it disadvantages the traditional Mediterranean diet. For this reason, the Italian government has proposed an alternative to the European Commission: a nutritional value labelling system known as NutrInform. This label makes use of a 'battery' symbol that informs consumers about the product's nutritional contribution towards their total daily requirements. In this respect, it is similar to the GDA/RIs system that indicates the recommended daily intake of nutrients. However, it is a less clear concept, the inefficiency of which (according to the advocates of the Nutri-Score) has been demonstrated in numerous studies.

A PERSONALISED ALTERNATIVE TO THE NUTRI-SCORE

What about a score that assesses food on the basis of your personal dietary preferences and health criteria? That is the solution that Carrefour is introducing in collaboration with the American Innit platform. You can see it as a personalised alternative to the Nutri-Score. So how does it work? A scientific committee gives a processed food product a health score out of 100: 60% of the points are awarded for the nutritional content, 30% for the additives and 10% for organic origin. The retailer then offers its customers the opportunity to personalise this score: the points given by the committee are varied to reflect the personal profile of the shopper. To make this possible, the shoppers fill in a form detailing their dietary preferences: vegetarian, sporting, slimming, no pork, etc. They can also indicate whether they want products with less sugar and salt or more fibre. Last but not least, they can say what they like to eat and what not.

The products that more closely match the consumer's wishes are given a higher personalised score. Carrefour also proposes alternatives for the less healthy products that score well. Scores are only given for processed products and, consequently, not for unprocessed fruit, vegetables, meat and fish. Likewise, products such as oil, sugar, alcohol and baby food are not assessed. The retailer guarantees that the customer data will remain anonymous and will not be used for marketing or advertising purposes.

App becomes food coach

The Nutri-Score is not only appearing on more and more products in our supermarkets, but is also starting to lead a digital life. Many of the smartphone apps launched by the retailers now make it easy to find out the Nutri-Score of a product by scanning the barcode.

In addition, a number of independent apps are now available that inform consumers about the origin and nutritional value of food products. The pioneer in this field was Open Food Facts, a project that first saw the light of day in 2012. It is a free and open data bank of the ingredient lists of numerous (more than 850,000!) food products around the world. The information is kept up-to-date by volunteers, in a manner similar to Wikipedia.

The most well-known of these apps is probably Yuka, which originated in France but is steadily conquering the rest of the Western world. The app was launched in 2017, after one of its founders went in search of healthy baby food in a supermarket. Trying to decipher the product label and interpret the detailed information it contained proved almost impossible. Consequently, Yuka wants to help people to decode these labels, so that they understand what they are eating. Similar to the Nutri-Score, the app gives food products a score, in which 60% of the points are allocated on the basis of the nutritional value (such as sugar and salt content), 30% for the presence of additives and 10% for whether or not the product is organic. The result is a colour code, from green (good) to red (bad). All the data used by Yuka comes from the packaging itself, claim the founders of the product, and is registered by the users themselves. In this way, the app is kept constantly up-to-date and, where necessary, amended. An automatic correction system tracks down and eliminates any irregularities. Brand producers and retailers are also allowed to add their products to the app.

At the start of 2021, the Yuka database contained details of more than 1.5 million food products and half a million cosmetics products. In France, more than 20% of all shoppers use the app, which is now also active in eleven other countries and has more than 20 million users in total. Its impact on the food industry is huge: according to the company, 94% of its dedicated fans use the app to guide their purchasing behaviour. Retailers like Intermarché, Auchan or E. Leclerc, as well as major producers like Nestlé and Unilever, readily admit that they have adjusted the composition of their products to get a better Yuka score. At the same time, however, criticism is also starting to grow: some experts say that Yuka is unscientific and lacks transparency.

Another app that is a little more out of the ordinary is the SmartWithFood app, a spin-off of the Colruyt Group that aims to become a true lifestyle coach. Users can scan products to get information about their composition, possible allergies, the Nutri-Score, etc., but they are also given additional advice on the basis of their personal food profile. This further allows the app to offer personalised recipes and inspiration tips, with the possibility to plan a week menu, which customers can then link to their Colruyt shopping list. In due course, it should even be possible to phone a food coach, who can give guidance about how to live a healthier life in your particular circumstances. SmartWithFood focuses on the positive evolution of eating habits and wants to collaborate closely with the food industry and food brands, so that it can grow to become a complete platform, valued by all.

The Dutch supermarket chain Jumbo has also developed its own app, the Jumbo Foodcoach, which wishes to help Jumbo customers to make conscious, responsible and tasty food choices. The app offers a wide range of varied and easy-to-make recipes and adds the required ingredients to a shopping list. This makes it easier when shoppers visit the supermarket or want to have the products delivered to their home.

The app was originally conceived for top sportsmen and women. Amongst others, the PSV Eindhoven football team and the Jumbo-Visma cycling team make use of it. However, the supermarket chain now wishes to make it available to a wider public. The roll-out for customers will take place in phases. In the first instance, it will be tailored to the needs of fanatical cyclists. Later, it will be adapted for other sports. Ultimately, it is intended to help everyone who is interested in good food and good health.

'More and more consumers are deliberately opting for a healthier lifestyle and want to feel good about themselves and their body. Among other things, this means opting for healthy food that is tasty and easy to make,' says CCO Colette

Cloosterman-van Eerd. 'Jumbo wants to help its customers to make these choices and aims to make delicious and nutritious food available to everyone. By offering a varied and inspired range, we invite people to make the right choices in different ways. At the same time, it also makes it more fun to shop at Jumbo.'

Even Foodmaker, a producer of 'healthy fast food', has developed an app that will help its users to optimise their eating patterns. The app suggests menus based on a very precise user profile, which not only takes account of age, gender and allergies, but also people's daily and weekly activities. For example, athletes who are preparing for a competition can get specific advice about their ideal diet. The company also offers to deliver the necessary ingredients or ready-made meals to their homes.

You can find out more about the Jumbo Foodcoach app here:

https://youtu.be/uGCu2I36bug

How 'nudging' helps people to eat (and live) more healthily: four inspiring examples

The term 'nudging' comes from the book *Nudge* by behavioural scientist Richard H. Thaler, who won the Nobel Prize for Economics for his research into psychological techniques that can help people to make better or healthier choices.

Although more and more consumers are concerned about the environment, these good intentions are not automatically translated into sustainable choices. Behavioural interventions to prompt them into action – Thaler's nudges – can help people to do the right thing.

Can food retailers take the lead in helping to gently push shoppers in a healthier direction? The risk, of course, is that this makes you seem patronising, so that you end up losing custom. But not if you can do it in a subtle and/or fun way.

A good example is provided by the Texan supermarket chain H-E-B, which is not blind to the serious health problems that exist in the Lone Star state: Texas has the highest rate of obesity in the US. Its answer was to create the 'Health at H-E-B' programme, which was intended in the first instance to set their own personnel on the right path. Internal workgroups were organised in which overweight employees encouraged each other to set and meet dietary objectives, which were then celebrated accordingly.

When it became clear that the concept seemed to work, H-E-B decided to extend it to its customers in a slightly amended form: healthier options were given better visibility on the supermarket's shelves and there were more special promotions offers for these healthier products. In addition, the chain organised the H-E-B Community Challenge, in which local communities were encouraged to compete with each other to see who would take the biggest steps towards a healthier lifestyle. This was followed by the Slim Down Showdown Challenge, which was targeted to weight loss. The results suggest that this playful competitive element seems to work. At the same time, H-E-B has used the opportunity presented by the obesity problem to strengthen its connection with both its staff and its shoppers.

Equally interesting is the experiment organised by the Environment Department of the Flemish Government in collaboration with the Colruyt supermarket chain, which was conducted over a nine-month period in 2018. By offering two smaller portion sizes of sausages, it was able to encourage roughly half of its customers to buy less. This led to a drop in sales (in terms of weight) of 18%. The customers actually bought as many of the smaller sausages as usual, but did not compensate for this by making other additional meat purchases. As a result, the objective of the portion reduction was achieved.

Similarly, offering vegetarian sandwich fillings alongside a non-vegetarian variant led to a 65% increase in sales of the vegetarian products. Specific references to the seasonal nature of some vegetables produced more complex results. Research showed that the intervention had little or no effect when consumers already associate a particular vegetable with a particular season (for example, pumpkins in autumn). In contrast, highlighting a weaker perceived link with a season (for example, cauliflowers in August) did have a stimulating effect on sales. The positive results were also maintained over time, which suggests that such 'nudges' can have a long-term effect.

Across the Atlantic, the Canadian retailer Loblaw, which runs both supermarkets and chemist stores, launched a smartphone app that offers customers personalised recommendations on the basis of their individual profile, in which they can define and set targets for their health needs. This is done in collaboration with health professionals, such as dieticians, who are available for consultation via live chat. The app is also linked to a loyalty programme: users can earn loyalty points if they perform their planned daily health activities and reach their pre-set targets.

The same idea forms the core of the Super Plus loyalty programme that the Delhaize supermarket chain launched in Belgium in the summer of 2020. If users of the app are willing to share their purchasing data with the retailer, they are given an extra 'health discount' on products with a Nutri-Score of A or B. The Nutri-Profile in the app shows how many products the customer has purchased and registers their Nutri-Score. The app also suggests healthier alternatives.

According to Ahold Delhaize boss Frans Muller, the programme is a big success: the number of users far exceeds initial expectations and Delhaize is winning market share. As a result, the retailer now plans to extend the same ideas to other chains in the group.

Getting older healthier, thanks to the right food

The market for healthy and appropriate food for senior citizens offers a huge potential for both manufacturers and retailers. By 2030, there will be 44 countries where more than 20% of the population will be older than 65 years of age. By 2050, Europe will have more than 150 million people over 65, says the World Health Organisation. This is a large, growing and relatively prosperous population group. Even so, for the time being the food brands have launched relatively few products aimed specifically at baby boomers and silver surfers. Marketeers currently seem to see more potential in millennials and Generation Z. But that is almost certain to change, since the opportunities in the seniors market are huge and increasing all the time.

People are living longer. Moreover, they are also staying healthy for longer. This means that they remain active. We are evolving away from the traditional life cycle of the past, when your retirement and pension (which will probably start at a later age) were simply a step towards the care home and death. The different phases of life are no longer inseparably linked with age. It is possible that in the 21st century we will move towards a cycle in which active people, after a brief 'pension pause', will embark on a second career until the age of roughly 80 years, when they will finally retire for good.

This will open up new markets in the fields of accommodation, health care, technology and, last but not least, food. People of an advanced age have special dietary needs and will be looking for food products that can help them to stay healthy and active for as long as possible. This might include food with supplements that strengthen natural immunity, or contribute towards stronger bones, joints and muscles, or better sight, digestion and heart function, or younger skin ... An example? Because older people produce less gastric acid, they absorb less vitamin B12 – necessary for healthy blood and nerves – from their food. Similarly, osteoporosis (brittle bones) can be caused by calcium and vitamin D deficiencies.

Some multinationals are beginning to develop this market, starting in the medical sphere. Reckitt Benckiser has already launched a wide range of nutritional supplements, including Move Free (a brand that supports the health of joints), Digestive Advantage (for improving the digestion) and Neuriva (for stimulating brain function). The Danone brand Nutricia has developed its Souvenaid range of dietary products for medicinal use during the early stages of Alzheimer's disease. Assist Black Coffee from Japan has been specially created to improve the concentration of older consumers. In the long term, we are also likely to see the emergence of more personalised food (see further in this chapter).

Having said that, there will also be plenty of opportunities in the broad food channel. Every supermarket has its shelves with baby food, diet food, allergy food, vegan food, etc. We regard this as perfectly normal. So why not add senior food to the list? They too have specific dietary needs and it would be short-sighted of manufacturers and retailers not to cater to them.

There is already a market for power food, enriched with extra protein and fibre, so that they offer sufficient nutrients but with smaller portions. This helps to combat a common problem with elderly people: undernourishment. As you get older, your appetite diminishes, so that you eat smaller and smaller portions. However, you have the same need for vitamins and fibre as younger people. Producers like Fortified and Huuskes (with its Chef Vitaal brand) already offer an extensive range of ready-made meals, to which extra vitamins, proteins and other substances are added in a flavourless layer. This kind of enriched food is not yet generally available in supermarkets, but is now common in many care homes and can be ordered via specialised catering companies.

This can also work to the advantage of companies that offer home delivery services for meal boxes or healthy instant meals. Older generations are becoming increasingly comfortable with the idea of online shopping – which sounds like good news for HelloFresh and Deliveroo!

Europe's problem with functional food

Discussion of products suitable for seniors brings us into the domain of functional foods, which are also known as nutraceuticals. This was a concept that originated in the 1980s in Japan, where the government found itself faced with the rising cost of health care. To tackle this problem, a regulatory framework was developed to approve certain foodstuffs which had been shown to have positive health benefits.

Even today, Japan still has the world's most developed market for functional food. That being said, there is no clear and generally accepted definition of what the term precisely means. It is more a marketing term than a scientific concept (Henry, 2010). In the Land of the Rising Sun, functional food is defined as food for specific health purposes; in other words, food containing functional components that beneficially influence the structure or working mechanisms of the body and are therefore used to promote improved health, such as supporting the digestive system, reducing blood pressure, reducing cholesterol, etc.

Functional food does not always need to be high-tech food. Some unprocessed, natural foodstuffs can also be functional. Think, for example, of carrots, which are rich in beta-carotene (an anti-oxidant) and help to improve your eyesight. Sometimes fruit and vegetables are 'enhanced', in order to strengthen the effect of a functional component. One example is tomatoes with a high concentration of lycopene, another anti-oxidant. The British Sainsbury's supermarket chain sells what it calls 'biofortified' foods, such as super-chestnut mushrooms, enriched with vitamin D and B12, or salmon cultivated on a diet that increases their content of omega-3 fatty acids. In this way, according to the retailer, food can become a means to pro-actively prevent chronic illnesses.

Conventional food products can sometimes also contain active ingredients that improve health. Breakfast cereals based on oat bran are a case in point, since this bran contains beta-glucans, which are large, sugar-like molecules that can support the reduction of cholesterol.

Other products have ingredients deliberately added by the manufacturer to provide specific health benefits. One of the most well-known examples is yoghurt drinks and margarines to which plant sterols are added, again to bring about a reduction in blood cholesterol levels. Familiar household brands such as Danacol, Benecol and Becel ProActiv have been in our supermarkets for years.

Equally well-known are the probiotic dairy drinks containing live bacteria that supposedly improve the working of the intestinal flora. However, these probiotics continue to be a source of controversy and the EU has refuted some of the nutritional claims made by the manufacturers as long ago as 2006. For this reason, producers like Yakult and Danone (with Actimel and Activia) now confine themselves to a more general claim that their products 'contribute towards the normal functioning of our immune system' (Actimel) or 'contain billions of bacteria that reach your gut alive' (Yakult).

The European Food Safety Authority is strict when it comes to dealing with the health claims of food producers and this has put something of a brake on the development of the continent's functional food market. Of the many thousands of health claims made by food manufacturers, the EFSA has so far only given the green light to some 260. Moreover, submitting a claim application is complex, time-consuming and expensive.

To make matters worse, the EU also uses a very precise form of wording when referring to these claims, which makes them almost incomprehensible for the average consumer. For example, there has been endless discussion about the use of the term 'normal' instead of 'healthy'. Hopefully, improvement is on the way. A European research project entitled Health Claims Unpacked wants to help manufacturers and marketeers to communicate more effectively about the health benefits of food, so that consumers can make better informed choices.

'Beauty food': a step too far?

Credibility is essential for the success of functional food. As far as the products that stimulate the intestinal flora or lower cholesterol levels are concerned, consumers (in part, at least) seem to have few qualms in this respect. But that is not always the case.

One famous marketing flop was the launch in 2007 by Danone of a special yoghurt: Essensis. This yoghurt promised its users beautiful skin, with the baseline 'Essensis feeds your skin from the inside out'. The product was sold in bright pink pots and the dairy producer allocated it a lavish 9.2 million euro marketing budget, hoping for a return of around 100 million euros per annum.

But it was all to no avail. After just two years, Essensis was withdrawn from the market. According to marketing experts, Danone made several capital mistakes. Initially, the multinational focused on a target group of young women. When that failed to work, they switched the focus to older women, between the ages of 40 and 60 years. The product contained active ingredients such as omega-6 from borage oil, vitamin E and anti-oxidants from green tea. But to mask the flavour of these components, Danone also added lots of sugar. Last but not least, the product was just too expensive. The advertising recommended a six-week cure of two pots of yoghurt per day, which pushed the price up to over 50 euros!

Danone's rival Nestlé also brought a cosmetic food product to market. Glowelle was a drinking water enhanced with extra vitamins and anti-oxidants, which (it was claimed) would improve skin health and help to combat ageing. The product was not sold in supermarkets, but in chemist's stores. Like Essensis, it lasted just two years before being ditched.

Not discouraged by these failures, in 2012 Coca-Cola also decided to chance its arm in this difficult segment. In collaboration with the French pharmaceutical group Sanofi, it launched a range of wellness drinks under the brand name of Beautific Oenobiol. These drinks were supposed to strengthen hair and nails, improve skin, promote weight loss and give added vitality. However, the project got no further than a small test phase in 200 French pharmacies.

Conclusion? For the time being, cosmetic food or 'nutricosmetics' continues to be a marginal phenomenon in Europe and the US. It is confined to a limited range of nutritional supplements available from chemists. This is in contrast to Asia – and above all Japan – where these products have now broken into the wider consumer market.

The personalisation revolution

The ultimate holy grail of health food is personalised food: products and recipes that are tailor-made to the specific needs of individual consumers during the different phases in their life. It is a highly promising concept, but one that is still very much in its infancy, in part because it requires complex and in-depth scientific research.

Even so, it is expected that before too long technology will make it possible to develop a hyper-individual approach to physical and mental health. Amongst other things, it is likely that by 2030 we will be seeing more smart diets based on smart DNA measuring technology that gives consumers greater insight into what makes them and their biological needs so unique. This will allow companies to make food and drink that is most appropriate for the data the consumers choose to share, using new production systems like 3D printing.

This personalisation of the food industry is known as nutrigenomics, a new domain that combines genetics and food science. The basic idea is that the human genome contains valuable information about the needs of each individual's body. Because we are all genetically different, in ideal circumstances our diet should be personalised to reflect these differences, since this is the best way to ensure a long and healthy life. This means that you first need to have your DNA mapped, following which a smart app will tell you what food you should eat and what food you should avoid.

This implies that consumers will have two expectations of the resultant personalised shopping lists, recipes and meals. They will not only need to be adjusted to reflect the individual's preferences and tastes, but must also improve their personal physical and mental well-being. This latter aspect is becoming increasingly important. For a growing number of people, working at their mental health is now on a par with watching what they eat and taking enough exercise, all the more so since medical research has now shown how the different systems in the body work together and how, above all, the brain, the digestive system and our emotions are all linked.

According to the experts, by 2030 a series of portable devices and smartphone apps will also warn people in real time about hormonal changes in their body and the onset of stress. This will result in changes in how, when and where food and drink is sold, so that people can always have instant access to the right products that can safeguard their health and/or enhance their mood at every moment of the day.

Habit is one such app. The company provides you with a personalised dietary programme based on a DNA smear, a blood test, other physical parameters and a questionnaire about your behaviour. The app offers a summary of recommended food and drink, a daily nutrition guide and several personalised recipes. Users can also link their personal Habit account to their Fitbit, so that their programme can be amended to reflect the level of their physical activity.

Nestlé is also active in this field. XiaoAI, an AI dietary assistant in China, is a smart speaker that provides information about food and health in the shape of appropriate recipes and even suitable musical accompaniment. In Japan, the users of the Nestlé Wellness Ambassador app send Instagram photos of their meals, on the basis of which artificial intelligence then suggests adjustments to their lifestyle and recommends nutritional supplements. Some of these users order products worth more than 600 euros per annum via the app.

This shows that consumers are happy to spend money on their health, which therefore has the potential to become big business in the years ahead. A report by MarketsandMarkets estimates that the global turnover will already have reached 14 billion euros by 2025.

In the chapter on 3D printing, we already mentioned the remarkable Sushi Singularity restaurant in Tokyo, which serves its customers hyper-personalised, 3D printed sushi. London restaurant YO! Sushi did something similar in collaboration with DNAfit, a company that markets DNA test kits. Customers could send the restaurant a sample of their saliva in advance, in return for which they received a report which tells them whether or not they are lactose intolerant, have an omega-3 deficiency, etc. YO! Sushi then advised them about the most suitable dishes from their extensive menu. The motor behind this concept is the software of the Vita Mojo order platform, which makes it possible for the customers to make choices from the menu based on their DNAfit account. For the moment, this is really more about marketing than science, because we still know relatively little about the link between our genes and our nutritional requirements. By comparison, for example, research into food ingredients that can positively influence our intestinal health is at a much more advanced stage.

Would you like a virtual visit to Sushi Singularity?

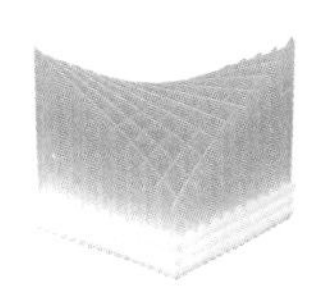

https://youtu.be/zNcfQrzMVTc

Gradations of personalisation

If, as seems likely, the trend for food personalisation continues to develop, this will present a serious challenge to the traditional food industry. This industry is currently geared exclusively for mass production; in other words, the same recipe for everyone. How can these giant multinationals switch to personal production: in other words, a different recipe for everyone? And how will they deliver these personalised products efficiently to millions and millions of individual consumers? Their business model will need to be turned on its head, but that is by no means self-evident. One interesting line of thought proposed by Oakland Innovation is to see personalisation as a spectrum, with different gradations.

The first stage would involve the development of products suitable for various, clearly defined target groups with specific needs. In keeping with the model adopted by G FUEL, an energy drink that claims to improve the focus and reaction speed of gamers, food producers could, for example, manufacture products that improve the intestinal health of target groups such as young mothers, infants or seniors. This a form of 'mass customisation', which is precisely what the Mars multinational is trying to achieve in collaboration with the start-up Foodspring, a manufacturer of sports food that has developed a digital coach that helps users to choose the products that are most suitable for them.

A second business model that has good personalisation potential is the HelloFresh model. In future, the company intends to provide more personalised meal kits. Because their current meal boxes already need to be filled with a series of separately packed ingredients, it is relatively simply to build in extra choice flexibility to provide different customers with different ingredients that are better suited to their individual health needs. In a similar vein, Amazon Fresh is working together with the personalised food platform Habit. The users fill in a questionnaire and submit blood test results for matters such as sugar, cholesterol, etc. Habit then provides them with a dietary programme and personalised recipes, while Amazon delivers the ingredients to their homes.

A third business model might be based on the approach of quick-service restaurant chains like Five Guys. They focus their activities on a single core competence (for example, burgers or tacos), which can be quickly and easily personalised by the addition of different selections of standard ingredients. Do you want a burger with no salad and extra ketchup? No problem! The question is, of course, how this model can be translated efficiently into a model for the production and distribution of pre-packed products.

The fourth and final model is the previously mentioned Vita Mojo model, which offers personalised food advice based on a nutritional profile.

The consumer is ready

Is this all just a pipe dream? There are three prerequisites for turning food personalisation into reality, according to Deloitte's Future of Food platform. The first of the three is greater insight into factors that influence the nutritional requirements of the individual. Science is already making giant strides in discovering how the intestinal microbiome, blood sugar level and even the amount and type of sleep can affect the way people react to food. Nevertheless, much greater progress still needs to be made.

The second condition is the development of the necessary technology to collect and process the mass of data that food personalisation involves. This technology is also evolving rapidly – think of fitness trackers and the advent of low-threshold solutions for blood, DNA and microbiome testing – but, once again, there is some way to go. We still need user-friendly apps that can combine personal food preferences and needs to give personalised dietary advice and help support food purchasing decisions.

Last but not least, consumers must also be ready to engage with the personalisation of their eating patterns. It would seem that this is increasingly becoming the case. Consumers are taking more and more account of the impact of food on their health. In this respect, it is interesting to note that a European study undertaken by Deloitte and Ahold Delhaize concluded that most consumers are now willing to share their data with retailers to a significant degree. This potentially gives the food industry the opportunity to develop services and business models based on this data.

However, the research also shows that while consumers have considerable confidence in the food retail sector, they are still holding back to some extent when it comes to personalisation. When asked whether or not they would be interested in personalised services that would help them to improve their health, the respondents showed most interest in recommendations for healthier alternative products (46%), followed by recommendations based on their past purchasing behaviour (45%). They were less interested in services that would give greater insight into their nutritional intake (37%), provide progress reports about their health objectives (37%), make health-related recommendations (32%), or compare their health objectives with those of other people (33%).

Deloitte see a number of different opportunities, such as personalised food advice based on food trackers, smart activity measuring systems and blood tests. Or precision and moment-appropriate food, which answers specific needs in relation to specific objectives. Someone training for a marathon would surely benefit from targeted dietary advice, wouldn't they? Time will tell.

HEALTHY DIGITAL APPLICATIONS

- Stance4HEALTH is an ambitious European project that seeks to promote better and more personalised food consumption by using smart mobile technology. The intention is to develop an e-diet app that will provide advice to its users based on their microbiome, eating preferences, lifestyle and budget. It will also be possible to link the app to a wearable device that monitors the nutritional composition of your body, your amount of movement, your hours of sleep and so much more. In a later phase Stance4HEALTH also hopes to develop individualised nutritional supplements and nutraceuticals for target groups like people who are overweight, or have coeliac disease, or an allergy to cow's milk.

- The Nutrition Wizard by Edamam quickly and easily provides an accurate nutritional analysis of a recipe, meal or ingredient. Users simply type in or speak in the recipe, meal or ingredient list and click on the 'analyse' button. The tool gives a complete nutritional evaluation in less than half a second, with details of all the relevant macro- and micronutrients, as well as labelling the food or recipes in accordance with more than 40 different kinds of diet, including all the most common allergy diets. People can quickly see how many calories a recipe contains and also how much fat, salt, sugar, vitamins and minerals.

- Day Two is a service developed by the Weizmann Institute of Science in Israel. It offers personalised nutritional advice based on a sample of your excreta, a blood test and a questionnaire. Using the resulting knowledge of your intestinal microbiome in combination with a number of other factors, Day Two selects food that will not cause your blood sugar level to rise. A related app provides users with personalised advice and recipes.

- Viome is another app that takes as its starting point an analysis of your microbiome and metabolism to recommend or discourage the use of certain foods. Viome now want to go a stage further, by marketing personalised capsules with nutritional supplements. To produce these capsules, the company has invested in a sophisticated and fully robotised factory.

- YOGUT ME is the first automatic yoghurt machine in the world, which allows you to make personalised yoghurt at home, packed full with intestinally friendly functional ingredients and free from all preservatives and additives. It works like a coffee machine with pods or capsules that contain different probiotics, prebiotics and other functional ingredients that can improve your digestive health.

- Eating too quickly leads to poor digestion and poor weight control. With this in mind, Slow Control has developed the HAPIfork, a digital fork that warns you by means of indicator lights and slight vibrations when you are eating too fast. The HAPIfork also measures how long it takes you to eat your meals, how many 'fork sessions' per minute you take and the length of the intervals between your 'fork portions'. There is also a related app with a coaching programme that is designed to improve your eating habits.

The components for a sustainable shift

The 20th century was an era of mass production and mass consumption. Although the century brought many people prosperity, it also created huge problems, including climate change and the destruction of the natural ecosystem. Today, we are seeing a change in mentality. We have a unique opportunity to anchor sustainability in new processes and to make the right choices for companies, consumers and the planet. Transparency is the new prerequisite.

Today's consumers demand sustainable products and solutions. Consequently, companies must make that sustainability a strong USP in their strategy. This must also go hand in hand with investment in social responsibility at the local level. The relationship of the business world with the planet and with society needs to be rethought. We need to understand and embrace the fact that together we form a single ecosystem. Companies that fail to take their responsibilities seriously will come under pressure: talent, customers, public opinion, legislators and activist shareholders will all turn against them.

Generation Z and the millennials, the most important purchasers of the future, are aware that they must pay the price for the errors of the capitalist period. The world paradigm will change dramatically, away from mass production and consumption, and towards the elimination of the general waste and inefficiency in society. It is time for a reset.

In their attempts to achieve climate neutrality in ten, twenty or thirty years, food retailers and producers must make the battle against food waste one of their main priorities. If we fail to act, 200 million tons of food in Europe alone will be lost by

2050; food that is produced, but not eaten. And that would leave behind a massive ecological footprint.

One of the remedies is better and smarter packaging. Because we need to be clear on this point: food without packaging is not a viable option. Even products that are offered for sale to consumers in bulk need to be properly packed on the long road that leads from the field to the factory and shop.

Packaging plays a vital role in the fight against food loss. It is essential for preserving food quality, safety and longevity, as well as having other functions in terms of logistics and marketing. However, today's packaging, and particularly plastic packaging, is a major source of pollution. The packaging of tomorrow will not only need to keep food in optimal condition, but also be circular.

A sustainable food system presupposes sustainable trading relations between all the different actors, from field to fork. This is where the need for transparency will face its greatest challenge. It is possible that technology will provide part of the solution but a change in mindset will also be needed, probably assisted by changes in legislation and procedures.

These are the sustainable options that we will explore in this chapter.

How packaging can become circular

The world is still addicted to plastic. The figures are mind-blowing. Each year some 8 million tons of plastic find their way into the world's oceans. More than 60% comes from packaging. 98% of the plastic packaging in Europe is still made from fossil fuels (oil). Just 14% is recycled. 40% ends up in a rubbish dump. More than 30% ends up in nature. It must be clear to everyone that this situation is no longer sustainable.

Europe is finally taking action. In 2018, the European Commission declared war against disposable plastic by, amongst other things, imposing restrictions on the sale of small utility articles, such as plastic knives, forks, spoons, plates and straws, but also things like plastic balloon sticks and cotton buds. The plastic producers now need to collaborate in awareness campaigns and also pay part of the cost of plastic waste management. In the future, they will be obliged to collect their waste plastic and recycle it. The EU's member states are already being encouraged to collect all their plastic bottles (or at least 90% of them). It is also possible that a compulsory deposit system for these bottles will be introduced.

Retailers, fast food chains and producers will be expected to ensure that less meal packaging and plastic cups are used. Europe is also introducing new labelling requirements, so that consumers will have a clearer idea about how much plastic packaging it contains, its ecological impact, and what to do with it once they have used it.

Industry is also now playing its part. More than 275 brands, retailers, recycling companies, government authorities and NGOs have undertaken to join the fight against plastic pollution. They signed the New Plastics Economy Global Commitment drawn up by the Ellen MacArthur Foundation and the United Nations. This agreement is important, because the signatories are responsible for 20% of the production of the world's plastic packaging and represent every link in the supply chain.

The core of the agreement is the commitment to end plastic pollution through the introduction of a circular approach. For the signatories, this means that plastic must never become waste. The Ellen MacArthur Foundation hopes to achieve this through a strategy based on six main pillars: (1) eliminate problematic or unnecessary plastic through innovation, new design and new methods of delivery; (2) reuse plastic whenever possible; (3) make all plastic packaging 100% reusable, recyclable or compostable; (4) ensure that the reuse, recycling or composting of all plastic packaging actually takes place; (5) separate the use of plastic completely from finite natural resources; (6) prohibit the use of dangerous chemicals in plastic packaging and respect the health, safety and rights of everyone concerned.

Almost all the top 100 companies (in terms of turnover) from the FMCG (Fast-Moving Consumer Goods) sector have now made pledges to take action to improve sustainability in the years ahead. Research by McKinsey has shown that these commitments focus on three key areas: full recyclability and the use of a higher degree of recycled material (60% of the commitments), a reduction in total plastic use (26%) and innovation in the use of packaging (14%).

To get the ball rolling, there is a lot of low-hanging fruit. Producers can start by eliminating unnecessary packaging, by designing packaging that is easier to recycle, by using mono-materials, etc. But after that, what are our best choices?

FIVE 'QUICK WINS' IN THE FIGHT AGAINST PLASTIC

In 2019, packaging producer DS Smith conducted a research study, which showed how European supermarkets could avoid the use of 70 billion plastic containers, equivalent to 1.5 million tons of plastic each year, if only they would opt for the use of renewable alternatives. The report highlights five categories in which quick gains could be made:

1. Shelf-ready plastic trays in the shelves of supermarkets are often overlooked, because customers do not take them home, but excellent cardboard alternatives are available.

2. In Europe, fresh products, such as soft fruit, are often packed in plastic trays, for which sustainable alternatives are available.

3. Almost all fresh drink units are packed in shrink film, even though cardboard and glue make more than acceptable alternatives.

4. Ready-made meals also make use of black plastic trays that are very difficult to recycle, a problem that can be avoided with the used of new cardboard alternatives.

5. Plastic trays and foils for meat, fish and cheese can also be replaced by renewable or recyclable alternatives.

Recycle or eliminate?

When the subject of plastic waste is discussed, accusing fingers are often very quickly pointed at the soft drinks industry – and in the first place at Coca-Cola. In fact, the company has some very ambitious climate plans. By 2030, Coca-Cola European Partners (CCEP) wants to reduce its emissions of greenhouse gases by 30%. By 2040, it plans to be climate neutral throughout the entire value chain.

These CCEP objectives are in line with the Paris Climate Agreement, which seeks to limit global warming to just 1.5°C. Part of the plan to achieve this target is the introduction of 100% recycled PET bottles. In some American states the soft drinks giant is already making use of these PET bottles (PET stands for polyethylene terephthalate). In Europe, Sweden, Norway and the Netherlands are leading the field.

Coca-Cola's colleagues at PepsiCo have similarly ambitious objectives. By 2022, the company hopes to make exclusive use of 100% recycled PET bottles in at least nine European countries. But PepsiCo is also exploring other possibilities. In 2018, the multinational acquired SodaStream, an Israeli company that markets a machine that makes it possible for people to make their own soda water and soft drinks at home. With its reusable bottles, SodaStream profiles itself as being a more environmentally friendly and healthier alternative to traditional bottled water and soft drinks in PET bottles. Moreover, it spreads this message via advertising campaigns that often border on the aggressive. A number of imitators have also brought similar concepts to the market, but the message is always the same: soft drinks producers produce above all plastic, whereas the real future is in refillable systems and reusable bottles.

This is also the line being taken by Mitte Water Lab, a German start-up that has developed a system that filters tap water and allows minerals and flavours to be added to it. According to CEO Moritz Waldstein, bottled water as we know it today will no longer exist in thirty years' time: 'There will always be a market for bottled water, but with a new emphasis on "on the go" packaging. The brands that remain will be premium brands, but the market will be a fraction of what it is today.'

Nestlé, an important water producer, is currently testing Refill+: another tap water converter, to which aromas, minerals and even caffeine can be added. The Belgian start-up Dripl has developed a soft drinks machine in which you can fill up your own water bottle with healthy biological fresh drinks mixed on the spot. It is already starting to make its appearance in offices and shops. Joi, Nooj Food and Modest Mylk are all producers of plant-based milk alternatives, but instead of offering a ready-made product they now sell a paste made from almonds or cashew nuts. All the customer needs to do is add water and mix. A jar of 425 grams makes seven litres of milk and also eliminates the need for litre-sized packaging.

Since 2020, there have been water fountains in three Delhaize supermarkets in Belgium. The company hopes that this environmentally friendly alternative to bottled water will save the use of 1.2 tons of plastic in each store each year. The initiative is a collaborative venture with water supplier Qanat, which provides both the filtered local water and the refillable bottles (PET or glass) that the customers use. The water has a slightly sweet taste and is available with and without fizz.

Is the disposable bottle reaching the end of the road? Notwithstanding all the good intentions mentioned above, the recycling of plastic packaging continues to be a sore point. In Europe, only about 40% of this packaging is recycled. For materials like cardboard (80%) or metal and glass (75-80%) the figures are much higher. In economically developing regions like Asia the situation is even more problematic, due to the increasing demand for packaging and the almost total absence of collection and recycling systems.

In Western countries, it is primarily 'on-the-go' packaging that finds its way into unsorted residual waste. The voices in favour of the payment of a deposit on disposable drink packaging (not just PET bottles, but also cans) are getting ever louder, because the voluntary sorting system is clearly not producing the desired results. In countries like Germany and Denmark a deposit system is already in operation, although in general the food retailers are not in favour of such systems because of the extra costs they involve.

According to critics, recycling is no more than a stop-gap measure that helps to maintain the fossil-plastic economy, whereas it is the dismantling of this economy that offers the best hope of a lasting solution. Moreover, recycling places the responsibility for action on the consumer, not on the industry. In this sense, the concept of upstream innovation offers better perspectives. This involves avoiding the creation of waste by exploiting innovation opportunities upstream in the value chain and implies greater collaboration between producers, logistics companies and retailers. In other words, tackling the problem at source.

This is very different from a downstream approach where the focus is first and foremost on consumers, as though it might be possible to solve the waste problem through better sorting and sympathetic 'clean-up' actions on beaches and roadside verges. The idea that a better environment starts with yourself has long since become outdated, says innovation strategist Stefaan Vandist in his recent book *Pretopia* (2021). Upstream innovation is no longer about greater awareness and behavioural improvement; it is about systemic change.

MIGROS

MIGROS COLLECTS PLASTIC

The Swiss supermarket chain Migros, in collaboration with a number of waste companies, has set up a project to encourage their customers to bring their plastic waste back to Migros stores in specially designed plastic collection sacks. The retailer then passes on all this returned plastic to recycling plants, which convert it into granules that can be used to make new packaging for Migros' own brands. In other words, it is a 'closed-loop' system, but one that also allows Migros to drastically reduce its use of plastic.

Food waste becomes food packaging

Notwithstanding its harmful environmental impact, plastic has excellent properties when it comes to the transportation and preservation of food. However, there is now an alternative for harmful and non-degradable plastic in the form of bioplastic that is made from natural materials, such as the starch found in potatoes, maize and sugar cane.

However, there is also a downside. Because maize and sugar cane are being increasingly used to manufacture bioplastics, this is putting increasing pressure on agricultural land. It therefore makes better sense to use this land for food production rather than plastic production. Moreover, many bioplastics are not biodegradable or suitable for home composting. So where does this leave the bioplastics industry?

One interesting possibility that is currently being explored is the manufacture of microbial and biodegradable polymers from food waste, derived from harvest residues and waste flows from the food production industry. These polymers can be used to make food packaging, thereby severing the food packaging sector's dependence on fossil fuels and ensuring that all the nutrients taken out of the ground are eventually returned there through composting. In this way, the entire process becomes circular, whilst simultaneously solving the dual problems of food loss and food waste (Guillard, 2018).

For example, an American research project has succeeded in making a packaging foil that is both edible and biodegradable. The basis for this new packaging is casein, which is a milk protein. The thin film of this substance holds oxygen five hundred times better than conventional films based on oil. Moreover, the development of the new foil opens up a number of other possible fields of application, precisely because it is edible. Imagine that you use it to pack freeze-dried soup. This would allow the packaging to be dissolved with the soup in hot water. Similarly, you could use it to pack spices or vitamins, which would again be released when the packaging is dissolved in hot water.

Another American project, this time by the Ecovative company, has found a way to make an alternative to environmentally unfriendly polystyrene, using organic waste and fungi. The biowaste is finely ground and then mixed with a special fungal culture (mycelium). The waste serves as food for the fungi, which start to grow. After five days the mixture is ground again, following which it can be given its final shape.

Packaging gets smart

Smart packaging not only protects the food products it contains, but can also analyse its freshness or ripeness, trace its origin and guarantee its safety. Time-temperature indicators monitor whether or not the parameters of the cold chain are respected during transportation. Sensors check the changing composition of the air (CO_2 and O_2) for packed fresh products and can sound the alarm when decay starts to set in or else signal when optimal ripeness has been reached. Smart anti-microbial coatings can combat the spread of harmful microbes and lengthen the storage life of the product. If you can combine all these properties with renewable materials, the result is intelligent, biobased food packaging that offers an answer to the challenges posed by carbon neutrality, food preservation and food safety.

The paper bottle offers a way forward

In collaboration with The Paper Bottle Company (Paboco), Coca-Cola and Carlsberg (amongst others) are testing the potential of bottles made from paper. These bottles currently consist of a paper outer shell with a recyclable plastic lining and screw top. The ultimate aim is to develop a bottle that can be completely recycled as paper. The bottle is suitable for both carbonated and non-carbonated drinks, but can also be used for cosmetic products and other fluids, which explains why L'Oréal is also showing an interest.

Coca-Cola was the first to test the new bottle in Hungary. For the moment, the cost is too high to be viable. Carlsberg sees paper bottles as a possible alternative for reusable glass bottles, which are heavy and therefore costly to transport in terms of energy.

A second skin replaces packaging

It is not just soft drinks that have plastic packaging. Fruit and vegetables are also sometimes packed in plastic, both to guarantee their freshness and to extend their shelf life. However, alternatives are now being developed for this type of packaging as well. Apeel Sciences uses the grape pulp left behind in wine presses to make an ultra-thin and edible coating that can protect fruit and vegetables from dehydration and oxidation. In other words, a kind of second skin or peel, which keeps food fresh for up to three times longer and is therefore an important weapon in the battle against food loss. Moreover, it also allows farmers to harvest their crops at a later stage of ripeness, which improves both flavour and nutritional value.

In the first instance, Apeel Sciences developed a coating for asparagus, a delicate vegetable that is usually transported by plane because of its limited storage period. In future, this transportation will be able to take place by ship, which is significantly cheaper. The second peel system is now also used on avocados. Del Monte is just one of the major fruit suppliers making use of this new technology, which has also found its way into supermarkets in the Netherlands (Jan Linders), Belgium (Colruyt), Norway and Sweden.

A tattoo for organic vegetables

In supermarkets, organic products like cucumbers, avocados or pumpkins are often wrapped in foil. This is because of regulations which state that organic products must be clearly distinguishable from 'conventional' products. And because the volumes of organic products are much smaller, it was decided that they should be the ones to be wrapped in plastic.

However, more and more smart retailers are finding ways to get around this problem. Organic vegetables are no longer wrapped in cling film, but are 'tattooed' instead! The tattoos – which are really more a kind of laser engraving – are part of the effort being made to ban environmentally unfriendly plastic from the organic assortment.

Initially, this 'natural-branding' technique was only suitable for vegetables with a thick skin like pumpkins, but now you can see tattoos on, for example, mangos, ginger and sweet potatoes. Moreover, the technique is perfectly safe: it makes no use of irradiation and adds nothing to the surface of the product. Consequently, it complies fully with EU regulations relating to the marketing of organic crops.

Bubbles replacing bottles

What if we could eat water instead of drinking it? Amongst other things, it would stop the use of millions of plastic bottles. Skipping Rocks Lab is an innovative sustainable packaging company, based in London. It is a pioneer in the use of natural materials extracted from plants and seaweed to create containers with a low environmental impact.

Their first product was called Ooho: a 'water bubble' inspired by the molecular cuisine of the Spanish chef Ferran Adrià. Brown seaweed extract and calcium chloride form a gel-like structure that acts as an elastic membrane and retains water inside. To protect it, a second membrane is added that functions as a hygienic container. Once you remove this 'outer skin', the remaining sphere is 100% edible. Or should we say 'drinkable'? Put it in your mouth, bite into it carefully and you can drink the liquid it contains! The outer membrane is not edible, but will biodegrade after four to six weeks. Moreover, this flexible spherical container can also be used for other liquids, including soft drinks, spirits and cosmetics, and is cheaper than plastic.

Scan the code for a fun demonstration of Ooho bubbles:

https://youtu.be/vVlwdUg3smQ

HOW PACKAGING WILL CHANGE IN THE FUTURE

- Packaging will be smart. Sensors will check if the product is still edible, even after its sell-by date. Or, conversely, will warn when a product has become inedible before its sell-by date; for example, because the cold chain was not properly respected.

- Packaging will be informative. Scan the QR code for recipes with the product you have bought and get a handy shopping list for the recipe ingredients you still need to buy.

- Packaging will be personal. Digital printing makes it easier to print packaging with your name or logo on it.

- Packaging will be reusable. Think, for example, of boxes that can be personalised and reused time after time for the delivery to your home of delicate items like fruit.

- Packaging will be edible. Your packaging will probably be made from plants, algae or proteins, so that you can eat them, feed them to your dog, mix them in your soup, etc.

The return of bulk products

Many food retailers are now testing out whole departments or even whole stores without packaging. This is a kind of 'back to the future' moment: at the end of the 19th century, nearly all food products were sold in bulk. Flour, sugar, coffee, salt, tobacco: people just brought their own pots and bottles to the shops and had them filled up.

The selling of food products in bulk has now once again become a familiar sight in specialised organic stores and is even making a return to some mainstream chains. Throughout Europe you can now find an increasing number of supermarkets and hypermarkets with bulk departments, where customers fill their own bags and pots from dispensers containing mainly organic products like nuts, seeds, legumes, breakfast cereals, pasta, etc. The most recent development is intelligent 'connected' bulk dispensers, which monitor their own supply and warn the retailer when a refill is necessary. Moreover, they can also be linked to an app via a scan code that gives customers details about the producer, the product's ingredients, possible allergies, etc. Nestlé is already testing this kind of digital bulk dispenser system for the sale of (amongst other things) instant coffee and cat food, complete with an app that provides all the information that would usually be on the package labelling.

The 'bring-your-own-box-or-bag' principle can also be used in delicatessen departments or at counters serving fresh meat and cheese. Marks & Spencer offers customers a discount if they bring their own boxes or jars. If you buy a ready-made lunch at their Market Place counter and provide your own reusable packaging, you get a full 25 pence (0.27 euros) off the asking price. Since 2018, customers have also received the same 25 pence discount if they bring their own mug or thermos flask for warm drinks.

Other food retailers are also moving in the same packaging-free direction. In Belgium, the Carrefour supermarket chain has been accepting the use of reusable bags and boxes at the service counters in its own hypermarkets and Market stores since the end of 2017. Because this trial was judged a success, it has been extended to Spain, where customers can now bring their own food containers for their purchases of fresh products – from fruit and veg to fish and meat – in more than a thousand Carrefour outlets. The only condition is that the containers must be sturdy, transparent and have a closable lid. Spar is doing exactly the same in Austria, where customers can bring their own pots and boxes for purchases from the delicatessen counter.

In particular, fruit and vegetable departments are ideally suited for replacing plastic packaging with bulk purchasing. In Europe, Ahold Delhaize has taken the lead.

In 2019, both Albert Heijn and Delhaize announced their ambition to make their fresh food departments as packaging-free as possible. At Albert Heijn, for example, they make use of the 'dry misting' technique, in which a fine spray of water is regularly applied to fruit and vegetables to keep them fresh for longer. Delhaize has made the switch either to bulk or to the use of recyclable packaging, such as cardboard. The retailer wants to cut the amount of plastic in its fruit and vegetable departments by 80%. A 100% reduction is not feasible, because some products continue to need a protective covering, either to keep them fresh or for logistical reasons. Where Delhaize cannot phase out packaging, it at least attempts to reduce it or takes compensatory measures elsewhere.

Major retailers are testing the packaging-free store

In Oxford, the British supermarket chain Waitrose is testing an innovative concept that has the potential in time to remove thousands and thousands of tons of plastic packaging from stores everywhere. Their pilot store in Botley Road makes use of a 'pick-and-mix' system for packaging-free deep-frozen fruit, serves wine and beer on tap, sells refills for cleaning products (in collaboration with Ecover), and uses dispensers for the sale of pasta, rice, cereals, legumes, dried fruits and seeds, not to mention freshly ground coffee that customers can take home in a reusable box. Most of the fresh fruit and vegetables have also had their plastic removed, as have the available flowers and plants. Shoppers can 'borrow' the necessary reusable boxes for a five pound deposit.

According to Waitrose, no supermarket has ever tested the packaging-free concept – launched under the name of Waitrose Unpacked – on such a large scale. And because it was deemed to be a success, it has since been rolled out in a number of other Waitrose outlets. The company's experts say that they expect to cut the use of one-off packaging by a massive 98%. Once the format has been further fine-tuned, it is planned to extend it to all Waitrose stores.

Waitrose's British rival ASDA also seem to see potential in the concept. They launched their own green test supermarket at Middleton (Leeds), with the aim of reducing plastic packaging as far as possible, while also stimulating the reuse and/or recycling of alternative packaging. The store has fifteen refillable dispensers for popular brands like Kellogg's, Quaker Oats, Lavazza and Taylors of Harrogate.

ASDA's private label pasta and rice are also available in bulk. Many other leading brand names and household products (Unilever, for example) are sold in refill packs. Where possible, fruit and vegetables are on display unpacked, which is the case (amongst others) for cauliflowers, mushrooms, apples, cabbages and cherry tomatoes. Plants and flowers are also sold unwrapped or else with a simple paper wrap-around.

To encourage its customers to shop more sustainably, the supermarket chain has launched a national price promise under the motto 'Greener at Asda Price', by which it guarantees that the price of its unpacked products will not be higher than the price of the packed equivalents.

Reinventing the deposit concept

Paying a deposit on food and drink containers has always been a contentious subject. A new and ambitious initiative in this field is Loop, a trial project involving prestigious names like Procter & Gamble, Unilever, Nestlé, Ferrero, PepsiCo, Coca-Cola, Tesco and Carrefour. It was launched in Paris and New York in 2019 and makes use of a concept devised by the TerraCycle recycling company. It can perhaps best be described as being a bit like the old 'milkman' system of days gone by, when fresh milk in bottles used to be delivered to your door each morning, with yesterday's used bottles being removed at the same time. This modern version makes use of a joint online platform, on which consumers can order groceries that are delivered to their homes in a special cool box. During the next delivery visit, the empty packaging is taken away for reuse or recycling. If they prefer, the customers can also collect their orders and return their 'empties' to physical stores in their neighbourhood.

To make the planned reuse feasible, the participating companies have redesigned some of their packaging. For example, a small steel bucket has been developed to allow the delivery of Häagen-Dazs ice-cream in New York, while in Paris PepsiCo now delivers Tropicana fruit juice in glass bottles and Quaker muesli in steel containers. Procter & Gamble has also created aluminium bottles for its Pantene shampoo and manual toothbrushes for which only the brush head (as with electric models) needs to be replaced. These used heads are then returned for recycling by Loop, as are used Pampers and sanitary towels.

In other words, the customers only buy the products, not their packaging. However, they do need to pay a deposit before they can receive the sustainable packaging in which the products are contained. This can amount to as much as three euros for decorated bottles and jars. An advanced IT system tracks the location of the various items of packaging and processes the payment and refund of the deposits. In the meantime, the project has now been extended to cities in the UK, Canada, Australia, Japan and Germany.

Founder Tom Szaky explains the idea behind his Loop project:

https://youtu.be/pY2rQfei2X8

SMARTIES IN RECYCLABLE PAPER PACKAGING

Smarties, a brand of the Nestlé multinational, is the first ever confectionery brand to make use of recyclable paper packaging. Nestlé has succeeded in developing a process that safely and viably recycles paper for the packaging tubes of this iconic sweet, which is popular with children around the world. To achieve this, Nestlé collaborated with the Institute for Packaging

Sciences in Lausanne. One of the consequences of their success was the need to make changes to the production line. The company has invested significant sums in a new confectionery factory in Hamburg.

'The transition of our Smarties packaging to recyclable paper is one of our most important sustainable packaging initiatives in the confectionery category. It is a further step towards achieving Nestlé's ambition to make all its packaging reusable or recyclable by 2025 and to reduce our use of plastic by one-third during the same period,' says Alexander von Maillot, global head of confectionery at Nestlé.

The new packaging is made of coated paper, with paper or cardboard labelling. After use, the tubes can be put out for collection with other waste paper. The packaging also gives information about recycling, to enhance public awareness. It is estimated that this initiative will avoid the use of 250 million pieces of plastic packaging each year.

The fight against food loss

If the food world wishes to do something about climate change, it needs to place the reduction of food waste and food loss high on the agenda. The current figures for both are mind-blowing. Wasted food is responsible for between 8 and 10% of all the greenhouse gases emitted. If food waste was a country, it would be the world's third biggest emitter of such gases after China and the United States. A quarter of the world's fresh water use is devoted to cultivating food that is never eaten._

The Food Waste Index Report of the United Nations Environment Programme (UNEP) contains the most complete details about the scale of food wastage that has so far been compiled, with data from 54 countries. The report looks at food waste in shops, restaurants and households and estimates that some 931 million tons, equivalent to 17% of all food production, is lost each year. Roughly 60% of this 'waste' is thrown away in people's homes.

To make matters worse, this figure takes no account of food that is lost earlier in the supply chain. According to the United Nations Food and Agriculture Organisation, food worth 363 billion euros is lost each year before it ever reaches the retail outlets. This is (almost) equivalent to 1 billion euros every day! The greatest losses of this kind occur in Southern Asia, North America and Europe. In Central and Southern Asia roughly one-fifth of all the food produced never makes it to the shops. The overall figure worldwide is 14%. Food loss puts unnecessary pressure on the environment and on the resources that are used to produce food in the first place. It wastes precious raw materials, creates pollution for no end result, and generates the unnecessary emission of greenhouse gases.

It is clear that action against food loss and food waste is urgently needed throughout the entire supply chain, from field to fork. In this chapter, we will look at a number of inspirational and promising initiatives taken by forward-thinking retailers and food manufacturers.

Best before ...?

One of the key sore points is the confusion relating to 'use by' and 'best before' dates. An awful lot of food that is still perfectly edible ends up in rubbish bins. The European Commission has estimated that this amounts to 10% of all food waste. The main reason for this is the fact that one in four of all consumers do not know the difference between the terms 'use by' (UB) and 'best before' (BB).

Perishable food that quickly degrades, like meat, fish, chicken, fresh dairy products, pre-cut vegetables and ready-made meals, will have a UB date. This is the last date on which the food in question can be eaten. After this date, the safety of the food is no longer guaranteed.

Food that is less perishable is given a BB date. This guarantees the quality of the (unopened product) until that date. After the date, the quality (taste, colour, aroma, etc.) of the product may decline, but it is still safe to eat. This is usually the case, for example, with canned food, pasta, rice and other dry foods that you can keep at room temperature. In other words, there is no reason to throw these products in the bin after the BB date – but that is precisely what many consumers do.

Awareness campaigns can help to correct this. Another possible solution is to think of a better (i.e. more readily understandable) term than 'best before'. With this in mind, the EU is planning a simplification of the rules and terminology relating to expiry dates. However, companies like Unilever, Carlsberg and Arla Foods are no longer willing to wait: they have already amended the expiry dates on a number of their products. In the Scandinavian countries, they have extended the 'best before' text by adding 'often good after'. In this way, they hope to prevent consumers from systematically throwing away perfectly good food.

In similar vein, Danone wants to make its contribution to the fight against food waste by replacing 'use by' with 'best before' dates on most of its products from 2021 onwards. With this change, the food giant wishes to indicate that many of its products are still perfectly safe to eat, even after the 'use by' date has expired. Analysis has shown that 85% of Danone products continue to be edible for a number of days without there being any problem of food safety. As a result, a whole series of brands, including Danone, Activia and Actimel, will henceforth have a maximum expiry date instead of a minimum expiry date. The remaining 15% of the multinational's products, such as puddings and fresh cheese, will continue to be labelled with a 'use by' date.

Some countries have introduced other initiatives to try to combat food waste. For example, in 2016 France became the first country in the world to prohibit supermarkets from throwing away unused/unsellable food. Retailers are obliged to make agreements with one or more social organisations to which they then donate the unsold products, as long as they are still fit for consumption. The system was an immediate success and the number of donations grew exponentially. An additional advantage is the fact that the legislation has encouraged retailers to keep tighter control of their stock management and labelling.

SMART INDICATORS AS AN ALTERNATIVE TO TRADITIONAL EXPIRY DATE LABELLING?

HelloFresh is also aware of the problems surrounding expiry dates. With this in mind, the company has started a collaboration with Wageningen University & Research. Working together, they have developed a smart shelf life indicator. The 'Keep-it' measures the temperature of a product in storage areas and calculates its residual usability. 'The results are promising and are currently being studied by the Ministry of Agriculture, Nature and Food Safety,' says Pauwel Wiertsema, CEO of HelloFresh Benelux. 'The aim is to see whether or not the Keep-it can be used to replace the classic system of expiry dates. This would be a big contribution to our ambition to deal more efficiently with food waste in Dutch and Belgian households.'

Let the app solve it?

One of the most interesting players in the fight against food loss is the smartphone app Too Good To Go, which was launched in Denmark in 2015 and is now well on the way to conquering the rest of the Western world. The start-up is currently active in fifteen or so European countries and in 2020 made the step across the Atlantic Ocean to the US. The app now has 33 million users worldwide, saves something like 100,000 meals per day and has become the largest global B2C marketplace for food surpluses.

The app combats food waste by helping restaurants and shops to sell their excess food instead of throwing it away. Catering outlets can upload details of the food they have over each day, so that registered consumers can buy it at a reduced price and then collect their fresh meals. The same applies for supermarket products that have a 'use by' date.

This is a clear win-win situation. The consumer can save food from the dustbin by paying a cheaper price for it, while the restaurateur reduces his amount of food waste, recovers part of his lost costs, enhances his 'green' image and comes into contact with new customers he might otherwise never have had. The app also helps retailers to quantify and analyse their food waste, allowing them to amend their purchasing and production processes accordingly. As a result, they learn how to better adjust supply to demand. Well-known supermarket and restaurant chains that have a structural partnership with Too Good To Go include Ahold Delhaize, Carrefour, Exki, Hilton, IKEA and Lidl.

This is already impressive, but the ultimate ambition of the app goes much further. Too Good To Go profiles itself as a movement, which not only wants to help companies and individual households to avoid waste, but also wishes to visit schools with an educational programme and even influence legislators to improve the existing food waste regulations.

From quick sale to happy hour

In effect, the app is a variation on the well-known supermarket principle of quick sales. Products that are approaching their expiry date are frequently sold with a discount of 30-35% for immediate consumption. But this system is far from foolproof: many of these quick sale products still end up in people's dustbins. In Belgium, the Netherlands and Luxemburg, Lidl offers fresh products such as meat, fish and cakes for the give-away price of half a euro on the morning of their expiry date. With this approach, Lidl hopes to cut its food waste by half by 2025. Half of the money the company earns with these sales is also donated to local food banks.

Albert Heijn is also experimenting with dynamic discounts for fresh products as a means to prevent wastage. They are carrying out tests at the poultry and fish counters in their supermarket in Zandvoort. An algorithm calculates the 'ideal discount', taking account of the condition of the products, the location, other bonus offers, the weather, historical sales records and the remaining level of supply in the store. The products are labelled with electronic price tickets that display two prices: the regular price and the reduced price in relation to a specific expiry date. This is more efficient and smarter than using price stickers.

The Finnish supermarket chain S-market uses a variation of this approach: it holds a 'happy hour' for food products that are approaching their expiry date. At nine o' clock in the evening, an hour before closing, the company lowers the price by 40% on hundreds of articles that are due to pass their 'use before' date at midnight. Often, the price on the products in question has already been lowered for a first time by 30%. This action is part of a concerted campaign to eliminate food waste. In addition to reducing surpluses, the initiative brings in new customers and helps people with a limited budget to eat better by offering them a greater variety of food at discounted prices.

DEALING BETTER WITH LOSS

Giving a standard discount on products that have nearly reached their expiry date? There are better ways for the retailers to deal with this situation, providing they learn to work with data-driven solutions. The food technology company Whywaste investigated best practices in thousands of supermarkets worldwide to gain the insights that allowed it to develop a solution that makes it possible for the retailers to adopt a smarter approach towards 'use by' and 'best before' dates. The resulting app monitors products that are approaching their expiry date and calculates the optimal sale price. In addition, it gives retailers a better analytical understanding of their loss figures, so that they can learn to avoid waste and increase profitability. The benefits are significant: 40% less waste and a reduction of 50% in the number of working hours that would otherwise be needed to check the dates on the products, according to Whywaste.

Wasteless is another similar solution. It calculates the ideal discount for products in relation to their remaining shelf life and rotation, so that it encourages customers to buy the products that are nearest to their expiry date. The system makes use of the GS1 Data Bar, which also contains these dates. The Italian hypermarket chain Iper has tested the system and concluded that it did indeed have an influencing effect on customer purchasing. When just one standard price is given, customers buy the product that is furthest removed from its expiry date. But when the price of a product is reduced as its expiry date approaches, shoppers are more often inclined to opt for the discounted version.

Surpluses for sale

More and more retailers are seeing the war on waste as an opportunity to establish a unique position in the market, while also playing an important social role. In the Danish capital Copenhagen, a supermarket was opened in 2016, specialising in the sale of surplus food. Its name is Wefood and it is a place where everyone can buy products at between 30% and 50% of the normal price. The initiative was launched by a local NGO, Folkekirkens Nødhjælp. The target group is not limited to people with a low income: the supermarket also aims to attract customers who are interested in solving environmental problems. Wefood is kept running by volunteers, so that overhead costs are kept as low as possible. The income from sales is used to fund the projects of the organising NGO, which seeks to eradicate hunger in developing countries. In theory (and in practice), the company has no problem finding sufficient supplies for its shelves: Denmark is estimated to have an annual food surplus amounting to 700,000 tons, even though this amount has been reduced by a quarter during the last five years. Wefood works with Føtex, a leading Danish supermarket chain, and with importers of citrus fruits and vegetables. It is clear that the project also enjoys considerable social support. The necessary capital to start the initial Wefood store – 1 million Danish crowns or 134,000 euros – was quickly raised by crowdfunding. In the meantime, Wefood now has four stores in Denmark and one in Finland.

Charitable impulses are likewise the motor behind Daily Table, an initiative by entrepreneur Doug Rauch, the former chairman of the American supermarket chain Trader Joe's. In 2015, he opened a first Daily Table branch in Boston. Now there are three. The organisation sells healthy, fresh food that is approaching its expiry date at dumping prices. Some of the products come from surpluses; others are badly labelled rejects or donations from supportive businesses. Daily Table wants to offer poorer sections of the community a healthier alternative to the omnipresent and unhealthy fast food that you can find everywhere in the US. Rauch calls his stores the first non-profit supermarkets. In addition to 'best before' products, they also sell ready-made meals made from 'saved' ingredients. Rauch is urging the government to take a closer look at expiry date labelling: 'As a society, we need to be aware of what these codes actually mean. Each year we are throwing away millions of kilograms of food simply because of the "best before" and "use by" dates. This has nothing to do with food safety.'

A similar initiative, but 100% online, is the Portuguese web supermarket GoodAfter, founded in 2016 by a start-up in Porto. The e-tailer only sells food that is approaching or has even exceeded its expiry date but with a guarantee that the food is still safe. Its discounts are high, up to 70% in some cases. The company is now active in Spain as well as in Portugal and obtains its products from manufacturers, wholesalers, supermarkets and other local stores that wish get rid of surpluses, unsold seasonal products or brands that they are removing from their range. GoodAfter works with the approval of the Portuguese Food Safety Authority, which allows products that have passed their 'use by' date to be sold, providing food safety is not compromised. Since the outbreak of the corona crisis, the company's turnover has increased by 250%.

Cooking with leftovers

Just as stores are emerging that only sell products from food surpluses, so there are also restaurants that work exclusively with products that have nearly reached their expiry date or are in some other way not fully compliant with the legal requirements. A classic example is the Amsterdam restaurant Instock, which was founded in 2014 by four ex-employees of Albert Heijn. In essence, the restaurant cooks with food leftovers. In the beginning, they came to collect these leftovers every morning from the local supermarket in their electric 'rescue' cart. Nowadays, the restaurant works primarily with Albert Heijn suppliers, since these generally have larger residual flows.

Instock does not only serve meals, but also sells products made from the ingredients it recuperates from others. For example, it has a webshop that markets three special beers: Pieper Beer (made from recovered potatoes), Bammetjes Beer (made from old bread) and Boos Bier (made from recovered raspberries). Instock Granola is made from brewer's spent grains, a nutritious malt residue that is left over from brewing beer. The owners have also published two cookbooks, with which they hope to help consumers to cook in a more environmentally friendly way and to waste less food at home.

This kind of initiative is also starting to appear elsewhere, and in general they are enthusiastically received. In the UK, a pioneering role was played by the 'zero waste' restaurant Silo in Brighton. Chef Douglas McMaster makes use of techniques like fermentation and drying to turn recuperated products into delicious ingredients. In Helsinki, Nola (Finnish for 'zero') is a gastronomic restaurant that operates on strict ecological principles, while in Copenhagen Amass is another fine example of a restaurant that wants to let nothing go to waste and makes delicious dishes from so-called 'leftover' products, such as the stems of cauliflowers or the peels of lemons, from which they can make miso.

These are praiseworthy initiatives, but if you realise that an average restaurant creates 70,000 kilograms of food waste each year, there is still a lot of work to do!

Silo chef Douglas McMaster explains more about his approach:

https://youtu.be/vnPVUnjadn0

BY-PRODUCT OF COFFEE BECOMES A HIP ENERGY DRINK

Coffee beans are encased in a shell that looks like berries or cherries. You can often see these shiny red fruits in coffee advertisements. As you might expect, the coffee growers are mainly interested in the beans and often throw the fruit husks away. That is a great shame – and very wasteful. The fruits are actually also full of flavour and you can use them to make *cascara*, which is tea-like drink. In coffee producing parts of the world, *cascara* is already well known, but it has never broken through into the mainstream. Until now. Things are starting to change, albeit gradually. Starbucks has already marketed a seasonal *cascara latte*. Smaller brands like Peloton, Nomad, Caskai and Ice Cascara sell bottled and canned versions. In Australia, Nestlé has launched Nescafé Nativ, a lightly sparkling drink. Because it is a by-product of coffee, *cascara* also contains natural caffeine, which is a stimulant. Up To Good, a Californian start-up, is marketing a fizzy *cascara* as an energy drink that gives you a 'clear, sustainable and lasting boost!' Each can contains around 80 mg of caffeine, which is comparable with a shot of espresso. Another good thing about *cascara* is that it creates an extra source of income for the coffee growers, but one that is derived from what was previously a waste product that used to be thrown away. It is a good example of upcycling, which ensures that an edible part of a valuable crop does not remain unused.

CREATIVE INITIATIVES

- The Dutch food designer Chloé Rutzerveld has developed Strooop!, a plant-based syrup waffle made from waste products (more specifically, sub-standard carrots and beets) from the vegetable industry.
- Long-term unemployed men and women make fresh soup from food waste, which the Colruyt Group then sells under the brand name EnVie.
- Delhaize has also launched five different kinds of soup, made from vegetable waste and residues collected from farmers. In this way, the supermarket claims that it 'rescues' 90 tons of vegetables each year.
- The British market leader Tesco is once again selling unwashed potatoes. This allows the potatoes to be kept for longer, which is important, since potatoes are the vegetable that most goes to waste in British families.
- Kellogg's and the Seven Bro7hers Brewing Co have joined forces to brew beer made from rejected breakfast cereals: 'Throw Away IPA'.
- AB InBev makes crisps and popcorn from the cereal residues of its beer production. The brewery group collaborates with Protes, an American start-up that makes snacks with a high protein content.
- Apeel Sciences uses the grape pulp left behind in wine presses to make an ultra-thin coating that can be used to protect fruit and vegetables from oxidation. In other words, a kind of second skin, which can increase the shelf life of products like avocados by up to three times.
- Upprinting Food makes attractive and tasty products using food waste and 3D printer technology. For example, spoilt or overripe fruit and vege-

tables are mixed with stale bread to create purées that are baked into artistic-looking biscuits, which gastronomic restaurants use as garnishing for their dishes.

- The products of Genecis are not edible, but are certainly sustainable: they use food waste to make biodegradable plastics and other useful materials.

Retailers stimulate sustainable choices

The Eco-Score maps environmental impact

A climate score, which gives the ecological impact of an individual food product, would be a very useful aid for people who want to make more sustainable food purchases. Similarly, a label that indicated the level of greenhouse gas emission caused by the cultivation/manufacture of a food product would make consumers more aware of the environmental effects of the food they consume. Both tools could play an important part in helping people to make more environmentally friendly and sustainable choices.

However, it is easier said than done to make such an eco-label. Tesco, the biggest supermarket chain in the UK, abandoned its plans to detail the CO_2 footprint of all its products in 2016: the problem was just too complex and too few other retailers seemed inclined to follow Tesco's example. According to the company, it would have taken months to calculate the ecological footprint for each product, which even then would probably have been the subject of dispute.

However, the intervening years have seen some progress being made in this direction. After two years of development, a collection of French actors, including consumer apps (Yuka, Open Food Facts, Etiquettable), the online biomarket La Fourche, meal provider Food Chéri and the recipe site Marmiton, have launched an independent Eco-Score, which aims to make consumers more aware of the environmental impact of their food purchases. The label is based on factors that have an impact on the environment throughout the product's life cycle: from raw materials and agriculture, through processing and transport, to use and disposal or recycling. The Eco-Score also takes account of extra plus and minus points, such as sustainability labels, packaging, origin and impact on biodiversity. The assessment is based on data provided by Agribalyse, a programme of the French Agency for Environmental and Energy Management (ADEME). Each product is given an initial score of 100, which is then weighted on the basis of 'supplementary qualitative criteria' via a bonus-malus system, which takes into consideration factors like the recyclability of the packaging, the labels, the land of origin, season-relatedness, etc.

It is no coincidence that the Eco-Score bears a strong visual resemblance to the more well-known Nutri-Score, which indicates the healthiness of food products. Thanks to its five labels, from A (green) to E (red), shoppers can now see at a glance the environmental impact generated by a food product, allowing them to make comparisons between different food options. Some examples? New potatoes have an Eco-Score of A, as does low-fat yoghurt and basil. Liquid honey and eggs have a B score. Extra virgin olive oil is C, pure chocolate and chicken fillets are D, while smoked salmon and ground coffee are E.

In 2021, the Colruyt Group was the first major retailer to adopt the new Eco-Score. Via the SmartWithFood app, Colruyt customers can consult the Eco-Scores of 2,500 own brand products. Later, it is intended to also give scores for national brand products. Lidl likewise plans to test the label. In a first phase, it is conducting a consultation exercise with representatives from society and the political world. After that, a second phase will see the introduction of the label on selected food products in all of Lidl's Berlin supermarkets, to see how its customers respond. Depending on this response, the third phase would see the roll-out of the Eco-Score in all Lidl stores throughout Germany. The discounter has now extended the Eco-Score test to the Benelux region. The company says that it would also be willing to consider a sector-wide initiative.

Why discounter Penny is setting higher prices

Food is actually too cheap: many products should be twice as expensive, if the environmental damage they cause is taken into account. A kilo of organic mince should actually cost 20 euros, and not 9 euros, as is currently the case. A litre of milk would cost 1.75 euros, and not 79 cents. This 'real price' would take account not just of the cost associated with making the product, but also the cost of repairing the damage it causes to the environment in terms of carbon and nitrogen emissions, over-fertilisation, energy use, etc. The University of Augsburg calculated the 'real cost' for 16 of the house brand products of Penny, the discount supermarket of the Rewe Group. The retailer then displayed these 'inflated' prices alongside the current shelf prices in its new, sustainable branch in the Berlin district of Spandau.

The aim was to improve customer awareness of environmental matters by confronting them with the huge difference between the two prices. Using this logic, Gouda cheese should be 88% more expensive and mozzarella 52%. The differences are smaller for fruit and vegetables: bananas should be 19% more expensive, tomatoes 12% and apples 8%. In general, organic products also have smaller differences, with one notable exception: organic meat should be 126% more expensive, if all its real production costs were taken into account.

'We need to make the costs of our consumption clear,' says CEO Stefan Magel of Rewe. 'Only then will our customers be able to make well-informed purchasing choices.' The company sees this as a first step towards a solution for environmental problems. Of course, Penny customers are not expected to pay this 'real price'.

This is just one of the twenty sustainability initiatives that Penny has introduced in its Berlin store, which, amongst other things, is also equipped with an insect hotel. In the new Spandau branch, customers can find answers to questions like: 'How many products would still be on the shelves if bees and other pollinating insects no longer existed?' or 'How can I arrange my fridge in the correct way, so that my food stays fresher for longer?'

The fruit and vegetable department also sells 'imperfect' specimens, as a statement against food waste. Respeggt eggs guarantee that no male battery chicks were killed in the supply chain. Cleaning products are sold in recycled packaging. Labels on fresh dairy products inform customers that the products can still be edible for some time after the 'best before' date. 'Smell. Taste. Enjoy.' is the supermarket's advice. At the checkout, people are encouraged to make an extra voluntary donation to a local organisation that helps disadvantaged young people. 'Sustainability is increasingly becoming a decisive factor in the consumer's choice of where he shops,' explains Magel.

Towards transparency in the food chain with a single click?

If we want to achieve a new, healthier and more sustainable food system, this will necessarily involve close and fair collaboration throughout the entire food supply chain, from farm to fork. But it will not be easy. Today's food chain is highly complex and lacking in transparency. Manufacturers and retailers purchase the ingredients for their products from all over the world and there are a great many intermediaries in the supply chain, each of which needs to take a piece of the margin, often resulting in complaints that the cake is not equally divided.

When a French agricultural health insurance fund announced a few years ago that a third of all French farmers earned less than 350 euros a month, the story was widely reported in the media. Carrefour CEO Alexandre Bompard also mentioned this at the start of 2018, when he launched the group's new strategy, one of the main pillars of which is the role that retailers need to play in the food transition. This reference was not coincidental. Large supermarket chains like Carrefour attract the biggest share of blame for the inequalities in the food chain: it is they who form powerful international alliances to force down the prices for producers and manufacturers as low as possible.

Things are increasingly coming to a head. On several occasions in recent years the European Union district in Brussels has been blocked by tractors driven by angry farmers, demanding higher prices and a fairer EU subsidy policy. Similar protests have taken place in other European countries, but also in India and the United States. Farmers need to pay off the loans necessitated by their high investment costs with miniscule margins and often in circumstances of great uncertainty, where a few days of bad weather can potentially ruin a year's entire harvest. These same farmers are then forced to watch impassively as the supermarkets launch promotion campaign after promotion campaign for meat and milk at give-away prices. When they then see the profits that are being made by these mega-multinationals, is it any wonder that they get angry and ask what is the point of carrying on?

A gap between supply and demand

In 2020, an acrimonious conflict broke out in Germany between farmers and the Lidl supermarket chain about (too) low prices for pork. After the farmers had organised several blockades at Lidl's distribution centres, they were able to force the retailer to increase the price by one euro per kilogram. The German government is considering whether or not to introduce a tax on cheap meat to raise revenue to compensate the livestock farmers. Early in 2021, there was uproar in the UK when an agricultural organisation revealed that almost half of the pork in British supermarkets is imported cheaply, instead of giving priority to local farmers. In Belgium, it was the dairy farmers who protested when Albert Heijn cut the sale price of its own house brand of semi-skimmed milk, AH Basic, to just 55 cents. These are just a few examples of the many and often bitter disputes between the retailers and the farming world.

Is all this anger justified? According to Willy Baltussen, a Dutch agricultural economist at the University of Wageningen, the discussion needs a degree of nuance. 'Every link in the supply chain incurs costs, but they often also make a profit on what they produce. What's more, it seems as though that profit is more or less evenly distributed. There is no-one who earns an excessive amount, not even the supermarkets,' says Baltussen. He points to the gap between supply and demand as the main problem. 'This is a situation that has already persisted for some time. There is too much on the market. The supermarkets can sit back and relax, knowing that there is always going to be enough tomatoes or meat, so that they can wait for the lowest price.'

In other words, there are problems with the way the agricultural sector operates. Given the nature of their business model, it is very difficult for farmers to respond flexibly to changes in market demand. They decide far in advance what crops they are going to plant or how much milk they are going to produce. They provide generic products, not differentiated brands. This means that as suppliers they are easily interchangeable for the retailers, now more than ever in an era of worldwide competition. As a result, a fluctuation in consumption in China can have an impact on pork prices in the Benelux. To make matters worse, there does not seem to be any immediate solution on the horizon.

Retailer boycotts brand manufacturer

It is not only the farmers who feel disadvantaged. Food manufacturers great and small also complain about the massive purchasing power of the supermarket chains and even accuse them of an abuse of that power. They claim that the retailers, who sometimes block orders or remove brands from their shelves (or at least threaten to do so) if the manufacturers are not willing to adjust their prices downwards, are restricting market competition. In the long term, the manufacturers reason, this is not in the best interests of consumers.

In recent years, there have indeed been a number of serious confrontations between the supermarkets and the brand manufacturers. Many of these conflicts remain hidden from the outside world, since neither side is keen to communicate about them. Even so, some of them occasionally leak out. And when a number of large multinationals, including Pepsico, Nestlé, Mars, Douwe Egberts and Coca-Cola, were systematically targeted one by one with boycott actions during the period 2018 to 2020, it became clear to everyone that something orchestrated was going on. The products of the target companies were temporarily withdrawn from the shelves of Colruyt in Belgium, Edeka in Germany, Intermarché in France, Coop in Switzerland, Conad in Italy and Eroski in Spain. No-one had ever seen anything like it before.

The background? These supermarket chains – good for a combined annual turnover of 168 billion euros – had decided to join forces in AgeCore, a retail alliance that was intended to put them in a stronger position to negotiate with the major brand manufacturers. One of the alliance's founding principles was that the failure of a manufacturer to reach agreement with one of the alliance's members could have repercussions in the stores of the alliance's other members. The multinationals were quick to discover that this was not an idle threat ...

Purchasing alliances: essential but controversial

This kind of purchasing alliance is not uncommon and their compositions change regularly. The logic behind them is crystal clear. Food retailers are generally only active in a limited number of countries, whereas the multinational manufacturers are often active worldwide. Moreover, the margins of the manufacturers are up to ten times higher than those of the retailers. In these circumstances, the supermarkets say that they are powerless to resist unilaterally imposed price rises that in essence have nothing to do with the product or its production costs. In addition, the retailers also complain that the manufacturers operate a system of so-called 'territorial supply constraints', as a result of which prices for the same product can vary considerably within the European Union, whilst at the same time making it impossible for the retailers to place orders outside of their own country. For example, in 2019 AB InBev was ordered to pay a fine of 200 million euros because it had forced consumers in Belgium to pay too high a price for its Jupiler beer between 2009 and 2016. The brewers prevented Belgian retailers from ordering the same beer at a cheaper price from the Netherlands. This, decided the European Union, is against the principle of the free movement of goods.

A report commissioned by the European Parliament in 2020 concluded that purchasing alliances are not a source of unfair competition. 'Buying alliances between retailers have become an essential part of the supply chain for grocery goods. They make it possible to offer consumers lower prices for the foodstuffs and personal care products that they purchase on a daily basis,' explained EU Commissioner Margrethe Vestager. Nevertheless, some of the practices adopted by these alliances remain controversial. In France, the Intermarché supermarket chain was forced to defend itself in court against an accusation of an abuse of power via AgeCore. The alliance is said to have demanded excessive contributions from manufacturers, without offering anything substantial in return.

Retailers play Robin Hood

There are a number of arguments to support the idea that the retailers have a dominant position in the relationship with their suppliers. To begin with, there is the trend towards greater concentration in the food retail market. In many countries, the top three players control 75% of the market. And scale means purchasing power. In addition, the retailers have more control over the sales process (many purchasing decisions are made on the shop floor), possess more data about shopping behaviour (knowledge is power) and compete directly against the brand manufacturers with their own house brands. Moreover, a large retailer can sometimes represent up to 20% of the turnover of a manufacturer. In contrast, a large supplier will seldom represent more than 2 to 3% of the turnover of a retailer.

Even so, a closer look at the figures reveals something remarkable: the brand manufacturers outscore the retailers systematically when it comes to crucial indicators like net profit margin, return on investment, shareholder value and market capitalisation. In other words, if the retailers are so powerful, why is it that they fail to turn this power into economic and financial performance? The answer is that the retailers are unable to capitalise on the benefits they extract from their suppliers. Instead, they immediately pass on these benefits to their customers in the form of lower prices and promotions. This is what marketing professor Marcel Corstjens calls the Robin Hood syndrome, in his book *Penetration* (2015): the retailers 'steal' from the rich multinationals to give to the poor consumers.

Retailers may have a powerful position in their relationship with their suppliers, but they do not have that same dominance when it comes to their sectoral rivals. The competition in retail is fierce. Food shoppers are a very volatile public. They easily change from shop to shop, if they can get a better deal elsewhere. In these circumstances, price competition is the retailer's most obvious weapon. It is difficult for them to differentiate themselves in any other way and, given their high fixed costs (including property and personnel), they have to appeal to as broad a public as possible. If they try to make the difference with a differentiating service or innovation, their rivals will soon copy it.

This results in a 'race to the bottom', with huge pressure on prices and margins. All the links in the complex food supply chain attempt to pass on this pressure to each other. If the brand manufacturers are put under pressure by the retailers, they will put compensatory pressure on their own suppliers: producers, transporters, providers of raw materials and, ultimately, the farmers. And if the pressure gets too great, there is always the risk that something will go wrong: loss of quality, problems with animal welfare, social abuses, food fraud, etc. The urgency for a solution is clear.

LEAVE PRICE NEGOTIATION TO THE ROBOTS

The relationship between the retailers and their suppliers is highly ambiguous. They are doomed to work closely together, in order to serve the best interests of the consumer. They exchange data about shopper and consumer behaviour. They develop special actions together. Yet at the same time their own interests are often incompatible: they both want to maximise their profit, if necessary at the expense of each other and their trading partners. This leads to hard negotiations, where dubious practices are sometimes used. These negotiations are usually assumed to be for the benefit of customers, because in theory they should lead to lower prices. That is one of the reasons why the EU has no objections against the retailers' international purchasing alliances. But if buying together is acceptable, fixing prices together is not, because this would eliminate the customer benefits.

The problem, however, is that price setting is not only very complex – there are many factors that influence the cost structure of a supplier-customer relationship – but it is also conducted by ... people. This means that there are always non-rational factors at play. Even if the negotiating partners attempt to reach a win-win-situation, the emphasis is always on winning for themselves, and both sides are aware of this. As a consequence, mutual lack of trust leads to less than optimal results. Viewed objectively, there is a point at which both parties can achieve maximum benefit, but in practice that point is never reached. And that costs money.

According to some experts, the answer is to take the negotiations away from the buyers and key account managers and put them instead in the hands of ... machines. Let robots, driven by artificial intelligence, carry out the task. Robots – if provided with the right data – are better able to see and assess the wider picture, process complex data sets and make optimal proposals that will work to the advantage of all concerned. Why? Because they are not hindered by emotions.

Sounds like science fiction? Walmart is already taking the first steps. The US retailer now makes use of artificial intelligence to negotiate its prices with hundreds of suppliers. In particular, Walmart calls on the services of the AI tool developed by the Estonian company Pactum. According to its creators, this system not only leads to stronger and fairer agreements that are better for both sides, but it is also capable of dealing with thousands of negotiations at the same time, which yields huge efficiency gains. So will the purchasing director of tomorrow be a robot?

Will technology finally make transparency possible?

The problems are known. But what about the solutions? Transparency seems to be the key that can open the door to a more honest and more sustainable food system. We need price transparency, so that the costs and benefits for every link in the long and complex supply chain become clear. Using transparent data, it must be possible to optimise that chain and make price negotiations more objective. The European Commission wishes to play a leading role in this process within the framework of its 'Farm to Fork' strategy, which we discussed earlier.

We also need transparency about origins, ingredients, working conditions and the technology used to harvest, manufacture, package and transport food products. Transparency makes it possible to better assess risks, to more easily avoid infections, fraud or loss of quality, and to increase consumer confidence. The expectations about the role that technology can play in these matters are high. According to various reports, blockchain, the internet of things (IoT) and artificial intelligence will not only lead to an improved food system, but will also save tens of billions of euros each year. How?

Blockchain to the rescue

Blockchain: it is the new buzzword of the modern business world. But it is much more than just the technology behind cryptocurrencies like Bitcoin. It also has huge potential for revolutionising the food chain. In simple terms, blockchain is a kind of decentralised database of transactions – in financial terms, a ledger – in which every transaction is digitally certified without the intervention of a central and controlling third party. This information is stored in 'blocks'. Together, all the individual blocks form a blockchain. Existing blocks can no longer be amended and therefore represent an immutable record. In this way, blockchain ensures the transport of data in a safe, transparent and verifiable manner. The system is disruption-proof, because it is not dependent on central servers, and is almost (but not completely) 100% fraud-proof, because the information is decentralised, public and transparent.

Moreover, blockchain also makes it possible to conclude so-called 'smart contracts'. These in turn make it possible, for example, to adjust prices and conditions automatically when agreements change – or when the parties do not keep to what has been agreed. As a result, the contracting parties only pay for the services or products that they actually use, and this is verified automatically. In other words, no more small print, no mountain of invoices and credit notes, no human errors and no fraud. The cost saving is huge.

In the complex food chain, blockchain makes it possible to take giant steps forwards in terms of the traceability of products and therefore leads to greater transparency, food safety and fair trade. In time, this technology will change the face of trading relations. Just imagine the impact that smart contracts could have on the relationship between farmers and supermarkets, between coffee growers and coffee roasters, or between A-brand manufacturers and purchasing alliances. The system might even make possible direct trading between producer and consumer. It truly is a paradigm shift.

Tracing from field to fork

Let us begin with the challenge of traceability. A number of concrete projects are already up and running. For example, the French supermarket giant Carrefour makes the supply chain of certain product categories fully transparent thanks to the use of blockchain technology. By recording the production and processing conditions for every link in the supply chain in a decentralised databank, customers can follow the entire trajectory that each product has followed. In this way, they can quickly and easily find out everything they want to know about, say, the chicken on their plate.

The buyers of a chicken that comes from the Auvergne and has a Carrefour quality label can check precisely where the chicken was raised, what it was fed on, when and where it was slaughtered, etc. Every link in the supply chain (breeder, slaughterhouse, processor, transporter) is obliged by Carrefour to register its information in the unalterable and ultra-secure database that is the blockchain. By scanning the QR code on the product, the consumer is automatically given access to the relevant information. By working with a blockchain system in this way, Carrefour wants to show to the world that it is 100% transparent, because the company itself has no access to this information and can therefore hide or change nothing. The retailer is now taking steps to extend the use of the same technology to other quality chains, such as tomatoes, farmyard eggs, Rocamadour cheese, fresh milk or Norwegian salmon. The aim is to have all quality chains registered on the blockchain by the end of 2022.

One of the most important facilitators in this field is the IBM Food Trust Network. This was initially set up in order to remove growing concerns about food safety and to simplify food recalls. Since then, it has developed into a way to optimise the food chain by devoting greater attention to the origin and freshness of products, as well as focusing on the reduction of food waste. The American Walmart chain demands that all its vegetable suppliers become members of the IBM Food Trust Network, so that field to fork tracing can be guaranteed. In 2019, Walmart also launched a blockchain platform in China, in collaboration with PwC and VeChain. This application makes it possible within a matter of seconds to gain access to product information throughout the company's entire food supply chain, such as production and processing, transport, storage, etc.

The platform also facilitates interaction with customers via QR codes, which is intended to enrich the shopping experience and increase consumer trust. Because Chinese consumers have ever higher expectations with regard to food safety and quality, this digitalisation of information relating to the tracing of foodstuffs is a good way to strengthen the brand value of the food suppliers and boost public confidence in their products. The platform already has data on more than a hundred of these products, including fresh pork, chicken, vegetables, dry food and other products from the Sam's Club own brand (a subsidiary of Walmart).

Food producers like Dole and Unilever make use of blockchain technology to map the movement of their products in the supply chain better and more efficiently. Nestlé is testing the technology to improve the traceability of milk from New Zealand. The food giant also wishes in due course to use the blockchain to track its supplies of palm oil from Latin America. For this, it will collaborate with OpenSC, a blockchain platform developed for WWF Australia in association with BCG Digital Ventures. OpenSC can track individual products from producer to consumer, which will help to eliminate illegal, harmful or unethical products, whilst at the same time improving the transparency of the supply chain.

Albert Heijn is also keen not to miss the blockchain boat. In collaboration with its supplier Refresco, the Dutch supermarket chain has logged every step in the product journey of its own brand orange juice on the blockchain. This juice has a Rainforest Alliance sustainability label and every consumer can now verify in a transparent manner what this exactly means. Whoever scans the QR code can discover everything about the origin of the bottle they have in their hand, from where and when the orange was plucked to the nature of the flavour and colour tests carried out by Refresco. The blockchain application is also interactive. With its 'Like2Farmer' functionality, customers can send messages and photos of thanks to the Brazilian farmers who grow the oranges for Albert Heijn's sub-contractor, LDC Juice.

Opportunities and thresholds

Another domain where the blockchain can lead to significant improvements relates to product recalls. Sometimes, manufacturers and retailers are forced to ask for products that have already been sold to be returned, because there is some kind of problem. Glass has been discovered in a jar of jam, a mistake has been made with the 'use by' date, a bacterial infection has been discovered in some cheese, the product can cause allergic reactions but this has not been mentioned on the label, etc. This kind of recall action is complex, expensive and never completely water-tight. The producers need to know precisely which products and which codes are involved, and which not. This is where blockchain technology can help. It records these matters much more efficiently and allows it to be recovered much more quickly when something goes wrong. This latter aspect is crucial, because speed is of the essence in crisis situations: if the product is already in the hands of the consumer, it is already too late ...

This is all very promising, but there are still hurdles that need to be overcome. The quality of the available data will always be a crucial condition: as in every chain, it is only as good as its weakest link. If someone accidentally or deliberately inputs inaccurate data, it remains permanently locked into the blockchain. Control is another issue. When a retailer like Carrefour or Walmart introduces blockchain for a specific supply chain, it is clear who is in charge of the process and who holds the data, even if the transportation of that data is decentralised. But for other data chains questions can sometimes arise about who is actually controlling the chain. A facilitator like IBM? Food safety organisations? Other third parties?

Transparency is also a potential problem. It is possible that not all the players in the blockchain will be willing to give everyone access to all their data at all times. Information about costs, prices, etc. remains sensitive and can be crucial in negotiations with both suppliers and customers.

Even so, some observers are already thinking several stages further. In time, they argue, the blockchain must make it possible to set up platforms that will make the retail sector irrelevant. Smart contracts between producers and consumers will change the entire business model from a push model – in which the manufacturers offer their products to the customer via retailers – to a pull model – in which it is the consumer who indicates precisely what he/she wants. In this case, it will no longer be the retailers who determine the available product range; the platform would give consumers direct access to the entire range of every available product. This is certainly feasible in practical terms, because the resulting short chains without intermediaries would eliminate the need for the expensive storage of large supplies and would waste less time and fewer resources in moving those supplies through the chain. Are we heading towards a decentralised peer-to-peer system based on smart contracts to manage and process e-commerce orders? The first concrete attempts have not been a success, as we shall see in the chapter *Why food retailers are wrestling with e-commerce*. But the idea is unlikely to go away ...

Transparency is the ambition, explains Frank Yiannas,
Walmart's Vice President of Food Safety, in this video:

https://youtu.be/SV0KXBxSoio

The problem with fair trade

If the farmers of Europe have reasons enough to take to the streets in protest, the situation is far more acute for the farmers in the South. In rich countries, consumers cannot go a day without basic food products like coffee, chocolate and bananas. These are the so-called 'strategic' references that you can find in almost every Western shopping trolley and for which special promotion actions in the supermarkets are a weekly event. The pressure on prices for these goods is great and their supply chain is anything but transparent. Most coffee and cacao growers are scarcely able to earn a living income, while a handful of huge multinationals dominate the sector and make massive profits. They ask the poor farmers to invest in better cultivation techniques, but in return seldom give them the fair price for their harvests that would make this possible and secure their long-term future. This situation has reached the point where it is no longer sustainable. Not in the least because climate change has also added significant extra pressure to production. Some areas around the equator risk becoming uninhabitable and therefore also unsuited to coffee and cacao farming.

This problem is nothing new. Since the 1970s, when the first charity shops were opened in Europe, there has been a growing realisation that the operation of the free market is not sufficient to guarantee the farmers in developing lands a price for their products that actually covers their costs and would make it possible for them to invest in sustainable cultivation methods, as well as giving them a comfortable lifestyle. The idea behind fair trade is to pay these farmers a premium over and above the going market price, so that they and their labourers can enjoy safe working conditions and a decent standard of living, supported by projects to improve education, health and infrastructure.

A system of quality labels was developed to assure Western consumers that the products they buy were indeed traded fairly. The first of these labels, in 1988, was Max Havelaar, named after the novel by Multatuli, in which he described the fate of poor coffee farmers in the former Dutch colony of Indonesia. 1997 saw the foundation of Fairtrade International, or Fairtrade Labelling Organisations International e.V. (known for short as FLO-eV), as the overarching organ above the many national federations and marketing organisations that had come into being during the intervening period. An independent control body, FLO-Cert, now inspects and certifies the products of member traders on the basis of some 250 criteria.

Today, you can find products with the easily recognisable Fairtrade label and other quality labels for fair trade in almost every supermarket. And the number of such products and their turnover is growing each year. They include anonymous fresh products as well as private labels and well-known A-brands. In this sense, it is no exaggeration to speak of a great success.

However, this is still a large 'but ...': more than 30 years after the launch of the first fair trade quality label the situation for the farmers in the South has not significantly improved. Poverty, slavery and child labour have not been eliminated. Many growers and their families are still not able to live a dignified human existence on the price they are paid for their crop. In spite of its growth in recent years, the Fairtrade label's turnover of 7.7 billion dollars only represents a fraction of the total food exports from developing countries. It seems that the Fairtrade premium is either not enough or is not being used properly. A particular sore point is the fact that production companies who work with the organisation are not able to sell their full harvest at the

Fairtrade price. In the meantime, the farming population in the South is getting older and older: the average age of a coffee farmer is now 55 years. Young people see how hard this way of life can be and have no incentive to take it up.

There are no simple solutions, because the problem is a complex one. Global warming is making the situation worse and it is estimated that half of the land currently used for coffee cultivation will be unsuitable for that purpose in 2050. There is also a political dimension: many fair trade products come from conflict regions, where local rulers put their own interests before those of the ordinary people. To top it all, the corona pandemic has significantly disrupted the supply chains on which the farmers depend.

An inflation in the number of programmes and quality labels

One of the most noticeable trends in this sphere is the way in which the large multinationals and international food retailers are taking matters into their own hands and setting up their own programmes and quality labels in the name of fair trade. In 2017, the British supermarket chain Sainsbury's, which often claims to be the world's biggest seller of Fairtrade products, launched its own 'honest tea' label for its house brand Fairly Traded. The retailer said that it wanted to work more directly with tea farmers in Rwanda, Kenya and Malawi, so that it could give them more appropriate support. Even so, the decision caused something of an uproar, because it was seen as the first serious criticism of the working of Fairtrade International, although Sainsbury's was quick to clarify that theirs was a pilot project that built on the foundations already laid by the world's leading international fair trade organisation (Subramanian, 2019).

In the chocolate sector, this trend began even earlier. It was as long ago as 2012 that Mondelez International, manufacturer of the Milka and Côte d'Or chocolate brands, launched its Cocoa Life programme, which organises actions in six cacao-producing countries to transform the living standard of the farmers and create a sustainable cacao supply chain. The programme makes knowledge, training and advice available to local communities, so that they themselves can take the necessary measures to meet and overcome the challenges. Others in the sector were quick to follow Mondelez's example. Nestlé initiated its Cocoa Plan and now applies the UTZ standards of the international non-profit organisation Rainforest

Alliance. It also works with the same organisation for coffee. Forever Chocolate is the comparable plan of the chocolate giant Barry Callebaut, which aims to make sustainable chocolate the norm. Not wishing to be left behind, Mars has created its Cocoa for Generations programme, with the ambition of improving the lives of cacao farmers in West Africa.

In this way, these multinationals and retailers hope to accelerate the pace of change and have a more direct impact. However, they still define their own sustainability standards, without much consultation with the local farmers. As a result, some critics see a hidden agenda, claiming that the companies' main objective is to secure their own supply chain and increase the output of their producers. And indeed, when companies set up expensive programmes and labels as PR marketing instruments, there is a real danger of 'greenwashing'. Moreover, the unprecedented inflation in the number of fair trade quality labels risks confusing consumers, who will no longer be able to see the wood for the trees. The website ecolabelindex.com now lists details of 455 sustainability labels. Not exactly a model of transparency.

A living income

A concept that is coming increasingly to the fore in fair trade circles is the idea of a living income. A living income must make it possible for a family to enjoy a decent standard of living, which means that it must be significantly higher than the minimum wage, irrespective of the country. Fair trade certification does not guarantee a living income, but only a minimum wage, which is not enough. Currently, there is no formal certification system that ensures such an income, but there are at least a number of initiatives moving in the right direction.

One of them is the Beyond Chocolate initiative launched by the Belgian chocolate sector, which hopes to guarantee a living income for all the sector's cacao suppliers by 2030. At the moment, the suppliers only earn roughly a third of that income. Does this mean that chocolate will have to become three times more expensive? No. If you examine the supply chain for cacao, you will see that the cacao farmer only gets 11% of the sale price, with the trader taking 7% and the processer 6%, while the chocolate producer takes a whopping 37%, as does the retailer. In other words, it is the producers and the retailers who take the lion's share of the cake, and this is where change needs to be made. With a price rise of just 10 cents per chocolate bar it would be possible to double the income of the cacao farmers. Unfortunately, things are not that simple and a truly sustainable Belgian cacao sector still seems a long way off. The first annual report of Beyond Chocolate concludes that there is still no real transparency in the sector, and this is the crucial factor if further effective steps are to be taken.

The ice-cream brand, Ben & Jerry's, which is owned by the Unilever multinational, has supported the Fairtrade label for a number of years, and in 2020 decided to give its cacao farmers in the Ivory Coast an extra premium on top of the Fairtrade premium and the minimum price for cacao that the national government has set for all the farmers. The aim is to break the cycle of poverty. A living income can mean good accommodation, health care, clean water and education for the children, with a little something left over to deal with unexpected eventualities.

Even relatively small initiatives can have a significant impact. In 2020, Eosta, a Dutch importer of organic fruit and vegetables, launched what it called the world's first Living Wage product in Western supermarkets: the Living Wage mango grown by Zongo Adama in Burkina Faso. Customers are given the option of paying slightly more per kilo for their mangos, to ensure that the people working for Zongo can also earn a living wage. In April 2021, a second product was launched: Living Wage avocados grown by Anthony Ngugi in Kenya. This time the customers no longer have a choice: if they want to buy Anthony's avocados, they must pay a Living Wage price that is just 2 cents per kilo extra.

Find out more about the background of the Beyond Chocolate initiative here:

https://youtu.be/mfJHEU3mJG8

Can blockchain make fair trade a reality?

One of the most inspiring stories in the fair trade arena is Tony's Chocolonely, a Dutch brand that aims to make 'slave-free' chocolate. In order to achieve this aim, the company needs to be able to rely on a transparent supply chain from cacao bean to chocolate bar. With this in mind, Tony's makes use of the Bean Tracker, a software system constructed with ChainPoint technology. This is not exactly blockchain, but something similar. The Bean Tracker digitally records all data relating to Tony's chocolate, from the cacao cooperatives in Ghana and the Ivory Coast to the final production of the chocolate in Belgium. It is a fairly complex chain, but Tony's claims that it knows at any given moment what volume of beans is being shipped, which farmers were responsible for what proportion of beans in each container, and precisely what is being processed in Belgium and how. The brand wants to further develop the Bean Tracker into a scalable solution that can be used by the whole industry as a standard tool to implement and guarantee bean-to-bar traceability.

The first major player to adopt the system was Ahold Delhaize, which is committed to buying fully traceable cacao for its own brand Delicata and willingly pays a higher price to work with Tony's partner cooperatives in Ghana and the Ivory Coast. The fact that Tony's, as a fair trade pioneer, collaborates with the huge Barry Callebaut multinational for the production of its chocolate has attracted a good deal of criticism, but Tony's does not intend to sit safely in its niche. By working together with a chocolate giant, it wishes to show that its model can also be applied on a much larger scale. Because 100% slave-free chocolate is only possible, according to Tony's, if the entire industry is on board.

Tony's Chocolonely has also conducted tests with blockchain technology, in collaboration with the Accenture advice bureau. The results were encouraging, but at this stage provided no improvement in comparison with the user-friendly Bean Tracker. The main challenge is (and will continue to be) how you get physical data – like the data relating to sacks of cacao beans on which there is no barcode – uploaded onto your physical platform. It is possible that in time further developments in the fields of artificial intelligence (AI) and the internet of things (IoT) will provide a solution to these problems.

Another interesting case in the same area is a project undertaken by the Wageningen Centre for Development Innovation in collaboration with Haitian avocado and mango growers, who often lose income as a result of smuggling. It is hoped to put a stop to this through the use of blockchain technology (Lambrechts, 2021). The QR code that the farmers fix to their fruits contains a link to all the cumulative data from the entire supply chain. As a result, the consumer has access to information about this chain from A to Z. One of the remarkable aspects of the project is the fact that the farmers remain the owners of the fruit until they reach the supermarket shelves, whereas in a traditional chain the fruit is constantly changing hands. This puts the farmers in a stronger position when dealing with the chain's intermediaries (Lambrechts, 2018). Even so, the project's ultimate conclusion was that blockchain is not a miracle remedy. Some data continues to be subject to manipulation. In part, this can be counteracted by organising the input of data via technology: trackers, sensors, thermometers, etc. But there is still a long way to go.

In an ideal world, all trade should be fair trade. If, in future, we wish to continue enjoying our favourite comfort foods – a relaxing cup of tea, a delicious piece of chocolate, a tasty banana, avocado or mango, and so much more – we will need to be prepared to pay the correct – fair – price, so that the living and working conditions of the farmers and producers in developing countries can be significantly improved. For this to happen, the supply chain will need to become more transparent, with technology as the main tool. Action will also need to be taken to combat global warming effectively, since huge areas of land around the equator are at risk of becoming uninhabitable. And it is in precisely those areas that the beans for our indispensible morning cup of coffee are currently grown ...

A recipe for a new distribution model

In a rapidly changing world, the food retail sector, which until now has been an indispensable intermediary between the producers and the consumers, will need to take on a new role. Digitalisation makes it possible for the farmers and the producers to reach consumers directly in a more efficient way. This is an absolute game-changer, a new reality that will force traditional retailers to radically reinvent themselves. We do not simply need digital retail; we need retail for a digital world. A world that will never again be what it once was. The recent corona crisis has seen to that, turning the existing retail paradigms on their head.

For decades, increasing urbanisation has been one of the most defining demographic trends. But it seems as if the COVID pandemic is likely to bring this trend to a halt or at least move it in a different direction. The crisis has suddenly and unexpectedly changed our view of the world in which we want to live and work. This, in turn, will have an enormous impact on the retail world and, consequently, on the food retail world. The current concentration of industry, labour, education and distribution in cities is coming to an end.

Retail, as we know it today, developed in parallel with the industrial revolution in the second half of the 19th century. And until today, retail has continued to be a part of that same old industrial fabric. The choice of shop locations, shop design, logistical streams, hours of opening, revenue models, etc. were all interwoven into that traditional industrial world, which is now starting to crumble. Retailers will need to shed their industrial skin, since the old industrial paradigm is soon destined to become a thing of the past (Stephens, 2021).

Many people now realise that in future they will no longer need to live in or near big cities in order to study and/or make a career. In megacities like Paris and New York the post-COVID exodus has already begun. Those who can afford it are exchanging their apartment in the city for a house with a garden in the periphery, commuting in to the office twice a week for meetings with their colleagues. Digital nomads can work wherever they want. They emigrate to regions where the quality of life is better and the cost of living is lower.

If, after the pandemic, more and more people continue to work at distance, as seems likely, this will have a high impact on footfall in cities. Shops and catering businesses that until recently were able to rely each day on the arrival and departure at fixed hours of masses of commuters will see their number of potential customers drastically reduced. This will lead to a serious loss of turnover for the traders who traditionally established themselves on access roads or close to train stations. During the lockdowns of 2020 and 2021, the sales of convenience stores in city centres were badly hit. For catering outlets the effect was catastrophic, as it also was for shops and restaurants in airports. The appearance of our big cities looks set to change, which will have an impact on the suburbs as well. Property prices in city centres will probably fall, while those in the periphery will probably increase. The urban fringes will welcome a growing number of new residents and new businesses, heralding in a true renaissance for these areas.

Which food retailers are likely to emerge as winners from this unparalleled crisis? The sales figures for 2020 show that neighbourhood and convenience stores scored better than supermarkets and hypermarkets. Price perception and proximity also worked to the advantage of discounters like Aldi and Lidl. Above all, however, we saw an exponential growth for a number of new models. The turnover for meal providers like Deliveroo and Uber Eats skyrocketed. HelloFresh had its most successful year ever. Online grocery shopping finally made its big breakthrough, to such an extent that the logistical systems were scarcely able to cope. The big online players became even bigger and more dominant. Amazon, Alibaba, JD.com or – closer to home in the Benelux – bol.com all recorded spectacular growth. And although these retail giants are primarily active in non-food categories, they are increasingly moving into the potentially lucrative territory of fast moving consumer goods (FMCG).

Moreover, they are not only doing this online. If they want to break though completely into the food sector, this will not be possible without a significant physical footprint. With this in mind, Amazon has bought the American organic food chain Whole Foods, has opened checkout-free Amazon Go convenience stores, and is now expanding its Amazon Fresh concept of compact supermarkets, selling a growing range of fresh food in a digital environment with automatic payment systems. Likewise, Alibaba has developed the Freshippo concept in various formats, from large food halls through fresh markets to mini-sized local shops. JD.com has done something similar with 7Fresh. What all these store concepts have in common is the integration of physical and digital shopping, uniting offline and online at a single location. This is what we call 'new retail'. And it is the future.

Why the food retail landscape is splintering

For several decades, the supermarket has been the privileged channel for the sale of food, drink and other household products. Specialist stores, public markets and local producers selling their own goods were all pushed into the background or disappeared completely. One-stop shopping has been the success formula from the 1960s until the present day. In essence, the business model of the supermarket has hardly changed throughout that period. The barcode and the self-scan were the only innovations of note. But that is now starting to change: new business models are increasingly challenging the hegemony of the supermarkets and hypermarkets. These new models vary from small-scale, hyper-local farm markets to revolutionary e-commerce concepts based on sophisticated algorithms and blockchain technology. Retail is changing, because the world is changing.

Perhaps the greatest problem facing the supermarkets and hypermarkets is the fact that the idea of mass consumption is now a thing of the past. In the 1960s and 1970s, when the supermarkets and hypermarkets first appeared and had their glory years, Europe as a whole was still largely in growth mode. Prosperity was increasing and for many 'consumption' was the fulfilment of a life's dream. Moreover, society was still fairly homogenous. Marketeers were able to make use of the generalised concept of 'consumers', who all had the same broad wishes: strong brands, quality products, attractive prices and deals, etc.

Those days are now gone forever. Our modern society is fragmented, multi-coloured and heterogeneous. Differences in fields such as purchasing power, cultural background, culinary preferences, family composition, use of time, etc. have all grown exponentially. And since the breakthrough of the smartphone everyone has a complete shopping centre available to them in their handbag or back pocket, providing access to an endless range of products that can be compared for price, quality and origin.

Given this new context, how can supermarkets and hypermarkets, with their 'everything-under-one-roof' approach, offer an effective answer to these fragmented needs? An amateur cook who goes to a hypermarket because of its wide range of fresh products might find himself increasingly irritated by the promotion section, where the bargain-hunters are in hot pursuit of goods at the lowest possible prices. The conscious consumer who wishes to use the organic department might ultimately be put off by the 'in-your-face' advertising campaigns of the big industrial brands. The ready-made meal section might well offer dishes from around the world, but they will never have the same authenticity for people of different cultural backgrounds, who are searching for the true taste of home that they remember from their youth. By wanting to please everybody, the supermarkets and hypermarkets are at risk of pleasing increasingly fewer people. At the same time, their business model is such that it does not allow them to make too many clear-cut choices, because they need a huge range and a huge turnover to cover their equally huge fixed costs. This is a Catch-22 for which the retailers, notwithstanding a number of interesting tests and pilot projects, have not yet found an answer.

In the meantime, more and more new concepts and business models continue to appear, which are all nibbling away, little by little, at the 'food and drink' monopoly of the traditional, physical food stores. And this evolution has been even further accelerated by the corona crisis.

Disruption comes from the outside

Of course, the retailers and brand manufacturers are closely studying these new business models, inspired – or frightened? – by the example of the large e-commerce players. Giants like Amazon, Alibaba or Google are not only developing powerful webshops and marketplaces, but also impressive global ecosystems with their own delivery services, payment systems, streaming services, advertising networks, loyalty programmes, own brands and so much more. It is inevitable that these 'empires' will continue to erode the FMCG market.

The truly disruptive innovations often come from outside. Look at the car industry, for example. The disruption is not being caused by BMW or Renault, but by Google and Tesla. Amazon is more a technology company than a retailer, and is therefore not hindered by experience and tradition, so that the company is free to experiment as much as it wants with new delivery services and checkout-free stores. To actually earn his money, Jeff Bezos relies on the exploitation of cloud services and the sale of advertising. Newcomers like Ocado, Picnic or HelloFresh also profile themselves first and foremost as tech start-ups. Domino's calls itself 'a technology company disguised as a marketing company disguised as a pizza company'.

It is no coincidence that the FMCG multinationals are increasingly trying to buy innovation through participation in start-ups: Unilever bought the Dollar Shave Club, Walmart bought Jet.com, Carrefour organises hackathons and has entered into partnership with Google. Imitation is another strategy: traditional supermarkets are now also launching their own delivery services, meal boxes, apps, urban farms, etc. Are the copies as good as the originals? That remains to be seen. In the meantime, the new players and business models continue to eat away at the core business and turnover of the large supermarket chains, with corona adding extra fuel to the fire. The supermarkets will need to show a great deal more agility and flexibility, if they hope to emerge from this perfect storm as winners.

Inspiration, variation, ease

The breakthrough of meal boxes is a classic example of what is happening. The name HelloFresh has become a synonym for the concept whereby consumers have the ingredients for their meals delivered to their homes in suitable sized portions with clear preparation instructions. D-I-Y meals in a box! The advantages for the consumers are numerous: they do not need to go to the supermarket, they are provided with effort-free inspiration and variation, they learn about new ingredients and recipes, and they have the satisfaction of making their own balanced meals. And because everything is delivered in the right portion size, there is no waste! Convenience does not get any easier than this!

Moreover, HelloFresh also knows how to retain its customers. The figures achieved by the company in the corona year of 2020 were impressive: turnover more than doubled and net profit increased tenfold, thanks to the company's 5.3 million customers. Even so, the company remains a minuscule player in the massive food market. That being said, HelloFresh and others like it continue to take a growing number of bites out of the market share to the big supermarket chains. And because the chains are weighed down under the burden of their massive fixed costs, every bite hurts.

Meanwhile, the king of the meal boxes now has a number of imitators of its own. The American Blue Apron, Marley Spoon in Berlin and Mealhero in Ghent (a deep-frozen variant), amongst others, all hope to surf on the changing patterns of cooking behaviour. The retailers and brand manufacturers are playing as well. Carrefour, in collaboration with SimplyYou, now offers its own meal kit. Albert Heijn has done the same with its Allerhande (Everything You Need) box. Food multinational Nestlé has bought the British recipe sellers SimplyCook and Mindful Chef. Working in the opposite direction, HelloFresh is investigating the potential for selling its boxes in supermarkets in tests with Sainsbury's and Delhaize. It is looking increasingly likely that the meal box is more than just a hype: it is here to stay.

The Nespresso model

What makes HelloFresh and its fellow start-ups such a threat to the traditional supermarket world is the fact that they approach consumers directly. They no longer need the supermarket as an intermediary. As a result, they have become the owners of their own customer data. This data is worth its weight in gold, since it makes it possible in ideal circumstances to predict which customers will order which recipes. This offers huge potential gains in terms of efficiency, because you can predict your sales and therefore your purchasing requirements for ingredients much more accurately. In contrast, most brand manufacturers are bound hand and foot to their customers, the food retailers, who are seldom willing to share their customer data, unless for a serious sum of money.

For this reason, the holy grail for many manufacturers is the so-called Nespresso model. This coffee brand, which is owned by Nestlé, was originally intended for the B2B market, but quickly discovered that consumers were also interested in this kind of premium coffee for their own homes. In response, the company built up its own distribution channel almost like a fashion brand, with a mix of webshops and flagship stores. And to make purchasing as easy as possible, there is also a subscription system. Nespresso has a perfect knowledge of its customers and their preferences. Consequently, it has full control over the brand experience at every stage of the customer journey. Just as importantly, it no longer needs to negotiate with the demanding purchasing directors of the supermarket chains, although it does have to accept competition from other imitation coffee capsules at a lower price. A blue ocean never stays blue for very long.

The number of food brands that approach their customers directly is currently limited, but that number is growing. For the time being, the subscription model is finding it easier to break through in non-food; for example, razor blades (Dollar Shave Club), nappies (Ontex), animal food (The Farmer's Dog), etc. But change is on the way. A good example is Magic Spoon, an American brand of protein-rich, low-carb, sugar-free breakfast cereal for adults. 'We have searched for years for a healthy version of the addictively delicious sugar-bomb cereals of our youth, but we found nothing,' say cereal entrepreneurs Gabi Lewis and Greg Sewitz on their website. And so they developed a 'guilt-free' alternative of their own, which they sell online, but with the option to take out a subscription. The low-carb trend is not new, but so far Kellogg's have missed this particular train. Of course, the giant multinational always has the option to buy or imitate its promising start-up rivals.

The multinationals are waiting in the wings

Many of the successful direct-to-customer start-ups position themselves in niche markets like health food or diet products. Dirty Lemon is a D2C brand of healthy drinks. One of its most remarkable features is its use of a 'conversational SMS commerce platform' (c-commerce) to sell its products. These can be ordered via their webshop, but there is also a physical store in New York. This Drug Store is unmanned. The customers can buy what they want using their smartphone: they send an SMS to the number on the labels of the products they intend to buy and then simply take those products home. For the first purchase, the Dirty Lemon chatbot asks for the credit card number of the customer and for future purchases the customer's telephone number is used for identification purposes. Just 5% of the products taken out of the store remain unpaid, because the system is based on trust. The store also has a secret entrance at the back: if you have been in contact with the brand on a number of occasions, you are given an entrance voucher for a 'bar experience', where you can try out new Dirty Lemon drinks. The investors in this company? One of the main ones is Coca-Cola.

The French start-up Feed provides balanced meal alternatives and snacks throughout Europe. Its products are vegan, gluten-free, lactose-free and GMO-free. Customers with a subscription get a monthly package with a 20% discount, but the products are also available in around 5,000 stores in France, Belgium, Spain, Italy and the UK. The German start-up justspices.de sells herb and spice mixes online. Yooji delivers organic baby food to your home, an idea that Danone found interesting enough to take out a participation in the company. The Berlin-based Foodspring is specialised in functional food for 'a fitter, happier and more productive life'. Mars acquired the company in 2019 and added it to its Mars Edge division, which focuses on the development of promising nutritional solutions, including personalised food.

At the same time, some of the multinationals are now starting up their own test projects for direct sales to consumers. These are often small-scale projects, such as the wrapping of presents for Christmas and other celebrations. However, some schemes are more ambitious. Mars offers consumers the possibility to order personalised M&Ms for a birthday, baptism or wedding. In the middle of the corona crisis, PepsiCo launched two virtual 'sweets and snacks' platforms in the US. PantryShop.com and Snacks.com make it possible for American consumers to have all their favourite comfort food delivered free of charge to their home in just two days. You can also send snacks and sweets to your friends and family as a gift. In the Netherlands, the company (under the name Unwasted) also sells a number of short-term perishable products such as crisps and breakfast cereals direct to customers in 'surprise boxes'. The aim, as the name suggests, is to prevent waste. The boxes are available from the company's own webshop and via the anti-waste app Too Good To Go.

The farmers are also selling direct

Developments of this kind have also set the agricultural sector thinking. What if farmers no longer had to pass auction houses, wholesalers and supermarkets to reach their customers? Farmhouse stores are not new but they experienced a renaissance during the corona pandemic. However, they remain little more than small-scale and highly localised initiatives. But what if the concept could be up-scaled? E-commerce potentially opens up a number of interesting possibilities. For example, many farmers are also finding inspiration in the sharing economy. In almost every city players are emerging where you can buy vegetable packs directly from the farm on a subscription basis. The same is also possible for meat. In Belgium, the deeleenkoe.be concept (the name means 'share a cow') has been up and running for some time. Consumers find each other and join together online to buy a whole animal. You choose your preferred race and reserve your part of it. Once the entire animal has been sold, it is slaughtered, portioned and delivered to your home in accordance with your order. In similar vein, online farming markets are also starting to develop, where you can buy fresh, local and organic products. In 2020, the Brussels-based eFarmz.be saw its turnover triple and expects it to double again in 2021. In France, Kelbongoo has opened a series of city collection points for its farm produce. In the US, the Farmers Post start-up is developing an online concept whereby farmers sell their surpluses cheaply in boxes by post, so that consumers everywhere can have access to fresh, local products.

Ready-made meals brought to your door: the breakthrough of ghost kitchens

In the battle for the food and drink market, the food retailers have started to develop more and more home catering solutions, from ready-made meals to sushi corners in their supermarkets. In this way, the retailers are gradually impinging on what has traditionally been the territory of the catering industry. This trend, where retail and food service merge into each other, is known as 'blurring'. Moreover, it is a trend that works in both directions, with the domestic caterers attempting to persuade people that they no longer need to leave their homes to get a truly first-rate meal. And leading the way are the so-called ghost kitchens.

Take-away restaurants are nothing new (with Chinese and Indian as perhaps the best-known examples), but ghost kitchens are a much more recent innovation. These are restaurants without their own premises, who cook exclusively for collection and delivery services like Deliveroo, Takeaway or Uber Eats. In effect, these new platforms are the catering industry's answer to Spotify and Netflix. They cater (quite literally) to a whole new way of food consumption. With a ghost kitchen, consumers are just one click of their smartphone away from an entire meal for all the family. The ordering apps give a list of 'restaurants' that you can never actually visit. They are virtual brands, developed in response to a market demand. You just choose from their menus and the ghost kitchen makes the food and delivers it to your door.

Observers predict strong growth in this sector. According to research conducted by Euromonitor, the turnover for home-delivered meals more than doubled between 2014 and 2019, and more than half of the world's population no longer has a problem with ordering meals from a restaurant that they cannot physically visit. It is estimated that by 2030 the total turnover for the sector will exceed a billion dollars (880 billion euros), continuing the massive boost that was given to this type of eating during the corona crisis.

The huge advantage of a ghost kitchen is its lighter cost structure in comparison with a 'real' restaurant. Lower staff and property costs make this kind of 'dark kitchen' more viable, says Michael Schaefer of Euromonitor (Beckett, 2020). By way of illustration, he points out that 60% of the cost of a Starbucks Latte goes on renting premises and paying personnel.

Moreover, in this kind of kitchen it is possible to automate part of the production process. Within the next five to ten years, it will become possible to prepare dishes like pizzas, noodles, coffees, etc. fully automatically, which will not only speed up the production process but also lower the production cost. Even established names like McDonald's (in London) and Chick-fil-A (in Nashville) are now testing out the potential of ghost kitchens. The concept also offers possibilities for food brands to directly approach consumers in the same way.

The largest number of ghost kitchens is currently to be found in China, where there are more than 7,500. The market is also growing quickly in India (3,500) and the United States (1,500). In contrast, the market in the Benelux is still small. In Amsterdam, Albert Heijn is conducting a pilot project for a meal service under the name 'Allerhande Kookt'. A ghost kitchen prepares meals chosen from the extensive recipe database of Allerhande, which are then delivered to the customer's home by Thuisbezorgd.nl and Deliveroo. The first results suggest that this is not as easy as had been hoped. In Belgium, the Colruyt Group is also testing the dark kitchen concept in Brussels under the name of Rose Mary, which offers daily-fresh, ready-made meals. Whoever places an order on www.chefrosemary.be in the morning will have their meals bike-delivered to their home (or office) by evening.

Deliveroo is also investing in virtual kitchens. Under the name Deliveroo Editions, the company provides container kitchens to start-ups in the sector. These start-ups do not need to pay any rent, but Deliveroo takes a commission on their turnover. Customer data determines the locations at which the containers are positioned and also the recipes that will be prepared. The shows that data is the most important ingredient of all in these ghost kitchens. By using its customer statistics, Deliveroo gains perfect insight into the culinary preferences in each district. The battle for food looks set to continue – and it is by no means clear who will win.

THE VERTICAL APPROACH OR OUTSOURCING?

An interesting example in this area is the Berlin-based start-up Vertical Food, which currently runs two dark kitchens, where meals are prepared for six 'virtual' restaurants: Vadoli, Fresh's, Spyces, The Hummus Club, Royal Rolls and Spagettini. All six have their own ordering platform and social media pages, with the aim of catering to different kinds of customers more specifically. In this way, the company is less dependent on the larger ordering platforms and also keeps control over its own data. Vertical Food also keeps 'the last mile' in its own hands, making it something of an exception in a sector that is dominated by the larger platform players.

The New York ghost restaurant Zuul makes use of a very different business model. It does not wish to compete with existing physical restaurants, but instead seeks to profile itself as a 'ghost partner' in the background. The idea is that the growing popularity of delivery services is starting to weigh heavily on the capacity in restaurant kitchens. Zuul takes the preparation of these delivery meals off their hands. In this way, the restaurants can extend their circle of customers and their wider name recognition, but without the need to invest in additional kitchen infrastructure.

Meal delivery services can also deliver your groceries

In recent years, cycle delivery services have become a phenomenon in all our towns and cities. Today, Deliveroo, Just Eat Takeaway and Uber Eats are stock-listed giants with worldwide tentacles, although they are often under fire because of the way they require their couriers to accept an unprotected freelance status. In addition to their deliveries for restaurants, this sector has now also discovered the possibilities offered by retailers. If you can deliver meals to someone's door, you can also deliver groceries. More and more retailers are seeing the benefits of working together with cycle delivery teams, since it allows them to satisfy additional customers and purchasing moments without the need for major investment. This type of delivery is not suited to the larger 'weekly shop', but is ideal for emergency and/or impulse buys. Consumers now expect more flexibility and choice options. This service gives it to them.

Uber Eats delivers groceries in several European countries for (amongst others) Carrefour, Cora and Morrisons. Deliveroo has also concluded partnerships with Carrefour, Monoprix and Picard in France, Aldi in the UK, and Marks & Spencer in Hong Kong, Singapore and the United Arab Emirates. In the United States, personal shopping services like Instacart and DoorDash have flourished in recent years, and the general upward trend in the sector looks set to continue for the time being.

Another recent aspect of the same phenomenon is the emergence of so-called 'flash delivery': services that promise to deliver your groceries to your door within 10 to 15 minutes. As you might expect, this is typically a service for major cities. In 2020, Weezy was launched in London under the Belgian entrepreneur Kristof Van Beveren, who was previously the manager at the Showpad start-up. Weezy makes deliveries from local hubs within a quarter of an hour, either by electric scooter or bicycle. From London, the company hopes to expand to other British cities.

But Weezy will not have things all its own way. The Turkish flash delivery service Getir, founded back in 2015, not only wants to conquer the UK, but also the rest of Europe with its promise of delivery within 10 minutes. In Turkey the start-up already processes 5 million deliveries each month. Dija is another flash delivery specialist, founded by two former managers of Deliveroo. Their aim is to gain an edge over their rivals by making use of smart data to identify the purchasing patterns that will allow them to make their service more efficient. The German player in the field is Gorillas, which is already active in a considerable number of cities in their home country, the Netherlands and the UK. The Russian Yandex company promises delivery in 15 minutes from its own dark stores and now wishes to target Paris and London as its next objectives. It looks like the cycle lanes and pathways are going to be pretty crowded ...

Blurring or the battle for share of stomach

More and more supermarkets are testing food service concepts. The fashionable term used in Europe to describe this process is 'blurring': retail becomes food service and food service becomes retail, so that the difference between the two becomes harder to see. In the United States they use the term 'grocerant', a contraction of 'grocer' and 'restaurant'. Can in-store eating facilities help to decide the battle for food and drink – or share of stomach, as it is sometimes known – or is it just a transient gimmick? Consultants have been preaching for years that blurring is on the point of a major breakthrough, but in practice it is still largely a case of experiments in the margin.

Convenience

Blurring is certainly nothing new. The introduction of shops in petrol stations in the 1980s was soon followed by the development of 'food on the go' services on motorways and major urban access roads, aimed at busy mobile professionals who needed a quick snack and something to drink. As time passed, this offer became more sophisticated, but at the same time, while the food service companies were focusing on nomadic consumers, the food retailers were also starting to discover the potential of convenience eating and began to sell sandwiches, hot pastries, take-away meals, etc. In the Benelux, Delhaize Shop & Go and AH To Go were amongst those who led the way.

The logic behind the concept is clear: it aims to serve impulse-buy consumers at every possible moment of the day; in other words, whenever they want to eat, meaning 'now'! Shoppers no longer think in terms of channels. They expect the immediate and seamless satisfaction of their needs, whatever the time and whatever the place. Each meal or snack that the retailers can pinch from under the noses of their catering colleagues means additional turnover in a sector where margins are perilously small.

Share of stomach

This battle for the food and drink market is often referred to as the share of stomach. Supermarkets now realise that they are no longer solely competing with each other for market share, but also with every other outlet that sells food and drink products: the snack bar, the motorway service station, the company restaurant ... In this competitive struggle, every euro counts, which explains why the range of ready-made meals in supermarkets has exploded since the 1990s.

In recent years, another factor has been added to the food and drink equation: experience. By offering consumers a unique store experience, supermarkets hope to escape from the cut-throat business of pure price competition, whilst distinguishing themselves from their rivals. In addition, they also see customer experience as a weapon against e-commerce. You can't have a pleasant meal out on the internet, can you? This is the reason why shopping centres nowadays offer a much wider and more varied range of catering options. Even outside the food sector, retailers are adding food service elements to their operations, both to differentiate themselves and generate extra turnover. A hairdresser's with a coffee bar, a fashion store with a wine bar, a bike shop that serves sandwiches: they are all now familiar sights. And how many of us have never stopped to try a portion of köttbullar in IKEA?

Food service in the supermarket

Sushi stands are starting to appear in more and more supermarkets throughout Europe. Carrefour is opening Carrefour Cafés in its Belgian hypermarkets. And

Makro has been running self-service restaurants for as long as anyone can remember, although they are not integrated into the store concept.

In Woluwe, near Brussels, the luxury supermarket Rob (part of the Carrefour group) has a successful lunch restaurant, where chefs cook with exclusive products from the store's own shelves. CRU, the fresh market concept of the Colruyt Group, also has two locations with restaurants, operating under the name Cuit. But perhaps the most complete blurring concept is Delitraiteur, belonging to the Louis Delhaize Group. This formula combines a snack bar with a food offer that is arranged according to the different eating moments of the day: breakfast, aperitif, lunch, dinner, dessert, etc. Officially, these local stores fall under the rules for the catering sector, which allows them to have longer opening hours and a more flexible approach to the use of staff than an average supermarket.

In the Netherlands, the Jumbo Foodmarkt was the instigator of a blurring hype as long ago as 2013. This concept integrated different catering elements into the store setting, with a range of eating options that stretched from pizza and pasta to grills and Asian dishes. Shoppers could either eat on site or take their chosen meals home with them, in keeping with the company's 'make it, take it, eat it' philosophy. For example, in the pizzeria you could watch your pizza being made, eat it as soon as it was ready, or pack it up to eat later on, or buy the fresh ingredients that would allow you to make it yourself.

After the opening of the first branch in Breda, almost every Dutch supermarket felt itself more or less obliged to develop its own food service idea. In 2017, Albert Heijn launched a more fully-fledged concept. In fact, it is really two concepts in one: Deli Kitchen and Bakery Café. The Bakery Café is located at the entrance to the supermarket and sells sandwiches, rolls, juices and coffee (made from its Perla house brand). The Deli Kitchen is centrally positioned in the store and serves fresh meals ranging from salads to pizzas, grilled chicken and spare ribs. Customers can compose their own meals (in part, at least), which they can either eat on site or take home. In 2019, the concepts were renamed as the Allerhande Café and Allerhande Keuken, with the intention that Allerhande (which in Dutch means 'all kinds') should be the flag for all the company's catering operations.

Carrefour with lounge bar

It is not just in the Low Countries that food retailers are exploring the possibilities of blurring. The concept is catching on elsewhere in Europe as well. In Milan, Carrefour has opened a convenience store that combines a co-working place, a lounge bar and a restaurant with a modern mini-supermarket. The concept is known as Carrefour Express 'Urban Life' and as well as being handy for doing your grocery shopping is also popular with customers for a tasty breakfast, light lunch or evening aperitif. So perhaps we should be looking at it the other way around: not as a supermarket, but as a catering facility where you can also buy products.

The store is quite small: just 120 m² spread over two floors. On the ground floor, shoppers can visit the cafeteria for a wide selection of breakfast products, desserts and snacks. This is where they can also find a gourmet department, where you can order pizza, and a Tokyo Street bar, with freshly made sushi. As if this were not enough, there is also house-made ice-cream, take-away meals, a salad bar and a juice corner. Everything can either be consumed on site or taken home.

But the real innovation, according to Carrefour, is to be found on the first floor, in the co-working space with its lounge bar annex, where customers make use of the comfortable sofas to work, relax, eat, or enjoy one of the selection of 200 international beers. The Milan store is open every day from seven in the morning until ten in the evening and, in keeping with an excellent Italian tradition, there is a daily 'happy hour' (in fact, three 'happy hours') between six and nine o'clock.

American variants

Instead of 'blurring', the Americans prefer to refer to 'grocerants': supermarkets combined with a restaurant concept. The Whole Foods organic chain now has more than 30 stores with a full-service restaurant and 250 other branches with a quick-service concept, for which the chain works with external food service specialists like Frankies (Italian restaurants) and Genji Sushi.

Another chain, Hy-Vee, based in the Midwest, has dozens of supermarkets with a grill restaurant. The drugstore chain Walgreens runs Upmarket Cafés, where you can pick up a smoothie or some sushi. Wegmans, a family-run chain active in the eastern part of the country, likewise serves sushi, but also salads, pizzas and sandwiches in its Market Cafés. Some stores also have a Seafood Bar or a Burger Bar.

Great potential

Trendwatchers see huge potential in blurring concepts, not least because the generation of the millennials (and younger) have a different approach to food than their parents and grandparents. They cook less for themselves, but still regard food quality, authenticity and experience as being important. Casual dining concepts that focus on flavour and health are popular with this target group.

As a result, it is in the best interest of the supermarkets to make an effort to attract and satisfy at least some of this group's consumption moments. Whether they do this with a genuine in-store restaurant concept or through a good range of ready-made meals, salad bars, sushi stands, etc. is something that requires careful consideration. There are numerous different factors at play, such as salary costs, flexible working hours and specific legislative provisions relating to food safety and the serving of alcohol.

Why food retailers are wrestling with e-commerce

For many years, food experts were sceptical about the potential of e-commerce for meeting people's weekly shopping needs. In 2020, these experts were given a serious reality check. During the corona lockdowns, the online sale of groceries enjoyed exponential growth, which put all the existing shopping systems under huge strain. Although the supermarkets remained open almost everywhere, the demand for online time slots was so great that the maximum capacity was soon exceeded. Food retailers were suddenly required to take on more staff working longer hours to deal with the rush, while their suppliers also had to work flat out to keep the supermarket shelves and warehouses full.

At the peak of the corona crisis, online penetration in the UK rose from 8.1% to 12.4%. In France, the comparable figures were from 6% to 10.2%. In Italy, where e-commerce in general is lagging behind, the penetration level doubled to 4.3%, while in Germany, a market dominated by discounters without a webshop, the rise was from 1.5% to 2.9%. By the end of 2020, the online market share of grocery shopping was 2.4% in Belgium, 6.7% in the Netherlands, more than 8% in France and 14% in the UK.

These figures speak volumes, particularly if you take into account the fact that offline sales also peaked during the crisis, so that many food retailers had insufficient capacity to fully satisfy their increasing online demand. A survey conducted by Bain & Company amongst 7,500 European shoppers in May 2020 revealed that one out of every five respondents had unsuccessfully tried to place an online order in the preceding weeks. In other words, the online potential was actually much higher than the figures suggest. Bain also estimates that the online market will retain between 35% and 45% of its extra corona growth, as result of which the market has already reached the position that was predicted for 2025.

Paradoxically, this is bad news for the food retailers. E-commerce weighs heavily on their profitability. While 'ordinary' supermarket purchases have a margin of between 2% and 4%, online sales often result in a loss, because the retailers currently fail to pass on their costs sufficiently to the customer. The collection of orders from a 'dark store' is just about a break-even operation, but retailers who make home deliveries from their physical stores are sometimes faced with a negative margin of as much as 15%.

So why offer an online service at all? Because if they don't, they risk being knocked out of the game altogether. In other words, the food retailers find themselves between a rock and a hard place. What can they do? Bain suggests three possibilities. Investing in the optimisation of the supply chain is unavoidable. In particular, there is a need for centralised e-distribution centres and/or smaller local dark stores (possibly through the conversion of existing stores). Far-reaching automation can also significantly increase productivity. But both these investments are costly.

Searching for new sources of income is also an option. Possibilities include selling advertising space and banners in your webshop and app, offering new digital activation options to your FMCG suppliers, selling data, etc.

Last but not least, the food retailers need to think seriously about price setting and the passing on of (delivery) costs. Bain thinks that there is room for improvement. If meal delivery services like Uber Eats can pass on most of their costs to the consumer with little complaint, why should that same consumer's weekly groceries be delivered almost for free?

ORDERING YOUR GROCERIES? YOUR FRIDGE WILL DO THAT FOR YOU!

The fact that many household appliances are now connected with the internet opens the door for a whole new range of e-commerce possibilities. Printers that can order replacement ink cartridges and washing machines that can order washing powder refills no longer belong to the realms of science fiction. The same technology is now also making its entrance in the kitchen. Why not let your fridge (or cupboard or coffee machine) order what it needs when you are in danger of running out? When people make a manual shopping list, they usually manage to forget something. But machines don't forget: they get it right every time. This kind of smart fridge is already available for sale at the better electrical retailers. One such fridge is the Samsung Family Hub, which has a touch screen that takes over the functions of your smartphone, serves as a recipe index, and acts as a digital notice board. Thanks to three in-built cameras, you can look inside the fridge while you are out shopping, so that you can check, for example, whether or not you have still got enough eggs. The fridge can also send you a reminder when you run out of a particular product or its 'use by' date has expired. It can even suggest recipes with the ingredients you have still got, as well as sending through shopping lists to the delivery service of your choice! In Belgium, the Samsung Family Hub is already connected with Collect & Go, the delivery service of the Colruyt Group, while in the Netherlands it operates with the Albert Heijn webshop.

E-commerce in food: four business models

Food retailers and pure players approach e-commerce operations in different ways, depending on their business model and the maturity of their online division. Broadly speaking, there are two options for the preparation of online orders: in the (existing) stores or in so-called 'dark stores' (e-distribution centres). Similarly, there are two options for getting the orders to the consumers: delivery or collection. All the different combinations have advantages and disadvantages.

1. Preparing orders in the supermarket and having the customers collect them (click-and-collect)

A supermarket employee fetches the ordered goods from the store's shelves, puts them in a box, and the customer comes to collect everything at an agreed time. This method of working requires the least initial investment and is therefore a low-threshold way for supermarkets to start with e-commerce. But the operational cost is high: the goods in supermarket shelves are not arranged in a manner that facilitates efficient order picking. Moreover, once the number of online orders increases, the order pickers will start getting in the way of the store's offline customers and there will also be a problem of space to hold the orders while they are awaiting collection. When this happens, it is time to move on to the next phase.

2. Preparing orders in the supermarket and delivering them to the customers' homes

This method of working has more or less the same advantages and disadvantages as option 1. The food retailer can deliver the goods with its own staff, which is a serious additional cost that is not always passed on to the customer. Alternatively, the retailer can make use of an external delivery service. In many countries, the cycle delivery companies that originally focused on the catering sector are now also working with the retail sector. For example, Carrefour and Uber Eats have concluded a collaboration agreement for Europe. In the United States, personal shopper services like Instacart and Doordash (now a stock-listed company) have been delivering groceries to people's homes or offices for several years.

3. Preparing orders in a dark store for click-and-collect

To pick online orders more efficiently, retailers need to invest in specialised and (for preference) largely automated e-distribution centres. These centres are known as dark stores, because they are often converted from supermarkets that are unprofitable or sales space that is surplus to requirements. For example, hypermarkets frequently have surplus floor space that can be re-designated to supply their collection point (or a series of collection points in a region) more efficiently. This is one of the reasons why the 'Drive' model has become dominant in France, with efficient collection points in hypermarket car parks (although Drive does sometimes offer a smaller range of products than the physical stores). The most recent trend in this area is 'Drive Piéton' (*piéton* means 'pedestrian'), which makes use of city-based collection points, not on a hypermarket car park but in a convenience store in the city centre.

The main disadvantage? Dark stores require a large investment. The retailers also need to pass on the cost of transport from the e-distribution centre to the collection point.

4. Preparing orders in an e-distribution centre for home delivery

Large, robotised e-commerce distribution centres make possible the most efficient processing of online orders. The current leader in this field is the British pure player Ocado. This company began as an e-tailer, but now derives greater turnover and growth from selling its technology to other retailers, such as Kroger in the US and Marks & Spencer in the UK. These retailers make grateful use of Ocado's technical solutions, such as data centres, algorithms for the calculation of the most efficient delivery routes, a user-friendly interface in the app, and (above all) fully automated distribution centres, where robots now do all the work. In reality, Ocado is a technology company disguised as an online supermarket.

In general, most markets see delivery as the key to online success. However, it presents some serious challenges. The Dutch pure player Picnic uses electric buses travelling along fixed trajectories. Customers can track the current position of the bus to meet it for collection. For smaller orders in the cities, many retailers now use cycle delivery. Several players are also testing the viability of self-driving vehicles. The American start-up Robomart has developed a fully automatic, self-driving neighbourhood van that brings an assortment of up to 100 fresh products (vegetables, fruit, meat, cakes, etc.) on its cooled shelves to your door. It travels autonomously at 25 kilometres per hour over a maximum total distance of 130 kilometres and is already being tested by (amongst others) Stop & Shop, an American chain in the Ahold Delhaize group. At the High Tech Campus in Eindhoven, the Albert Heijn chain is also testing a fully automated delivery robot that answers to the name Aitonomi. Tesco and Metro are likewise experimenting with the self-driving delivery robots of Starship Technologies. The next step will undoubtedly be delivery drones, although it will still be some time before the technology and the necessary legislation are fully in order.

You can view a sample of Ocado's impressive technology here:

https://youtu.be/g2uMfpAHdGY

STILL NO REAL BREAKTHROUGH WITH VOICE TECHNOLOGY

In recent years, various food retailers, following the example of Amazon's Alexa, have added the possibility for customers to speak messages into their apps. Voice assistants certainly offer customers greater ease and convenience. They allow you to shop hands-free and it is faster than typing in a product or search task. Walmart, Carrefour, Jumbo and Colruyt all make use of Google Assistant for this purpose. Not only can customers make shopping lists with the system, but also search for recipes and get step-by-step instructions for their preparation. Even so, the voice assistants have not been universally successful. After two years, Albert Heijn pulled the plug on a test with Google Assistant, claiming that the speech technology was still not fit for purpose.

Blockchain to the rescue again? Not yet

The rise of e-commerce has also set the brand manufacturers thinking. Do we still need supermarkets if 80% of the weekly shopping of most consumers is perfectly predictable? Might it not be possible to automate these household purchases? If so, it would become theoretically feasible for the manufacturers to satisfy the consumers' needs for these repeat buys directly. But the manufacturers would need to work together, because it is not so easy as a producer to reach individual consumers. That is a result of the paradox of choice: customers expect the widest possible choice, but once they have it they experience choice stress! It is a delicate exercise for producers to find the right balance, which effectively means that they always need a curator. Direct-to-customer shops and webshops have so far only had limited success, because their ranges are generally too small.

So what is the answer? One possibility is the use of blockchain technology for the creation of an unhackable transaction register, where all relevant information can be stored safely in a peer-to-peer network. Blockchain is already raising the traceability of products to a higher level by recording every link in the supply chain in a decentralised databank. But from the perspective of the brand manufacturers, it has some other promising implications: the elimination of unnecessary intermediaries.

This, at least, was the dream of the Russian start-up INS Ecosystem, which had plans to cut the supermarkets (as an expensive intermediary between the producer and the consumer) out of the supply chain. The founders Peter Fedtsjenkov and Dimitri Zjoelin, who are also the brains behind the Instamart grocery delivery service, thought that their transparent online platform could make grocery shopping up to 30% cheaper.

The big advantage of an e-commerce platform via blockchain is that it works on the basis of a pull model. By removing the barriers in communication between the consumers and the producers, it becomes possible for the consumers to indicate their needs directly on the platform. As a result, it is no longer the retailers who determine the available assortment. Instead, the platform gives consumers direct access to the full range of all available products. Or that, at least, was the theory. In practical terms, it is certainly possible, since shorter chains with fewer intermediaries would obviate the need for large and expensive stocks, while also wasting less time and fewer resources. In INS Ecosystem's concept, a virtual currency was intended to play a central role: the INS token. Although customers would be able to pay with various currencies, all the marketing campaigns and reward programmes would be based on these INS tokens. At that time, this was a revolutionary idea. But then the Bitcoin exploded onto the scene ...

At first, things seemed promising for INS: FMCG giants like Unilever, Mars, FrieslandCampina and Reckitt Benckiser all signed collaborative agreements. In the Benelux, PostNL was lined up to do the deliveries. The postal company even launched its own grocery service under the name Stockon, based on a fixed delivery subscription for frequently used everyday products. Users of the app were sent a fixed shopping list every two weeks from which to choose, with the products being delivered to their homes by their usual PostNL service. None of the products in the distribution centre was the property of Stockon: the start-up worked with a 100% consignment system, so that it effectively functioned as a kind of platform for grocery goods. The collaboration seemed particularly interesting in terms of data collection, because Stockon shared the data relating to the products sold with its suppliers, which included Colruyt.

There is a good reason why all this is being written in the past tense: nothing more has been heard of INS Ecosystem in recent times. In March 2019, Stockon also threw in the towel. The company was unable to persuade sufficient external investors to finance its further growth. Just bad timing? Or the fault of a model that was unable to deliver what it had promised?

NOT AT HOME? NO PROBLEM!

Walmart is working together with smart box producer HomeValet to facilitate the delivery of perishable food products. This system has huge benefits for both the supermarkets and consumers. Walmart customers who sign up for the programme receive a HomeValet smart box. This has three temperature-controlled compartments, so that frozen, cooled and non-cooled products can be delivered and kept fresh. The delivery operative can communicate with the HomeValet to open it, so that he/she can leave behind the goods that have been ordered.

Customers order their groceries via the Walmart app and agree a delivery time. But the main advantage is that they do not then need to be at home to accept the delivery. All they have to do is leave the smartbox outside. What could be easier? For the retailer, this system also significantly expands their delivery options. 'While we don't have plans to do 24/7 delivery today, it certainly has a nice ring to it,' said a Walmart spokesperson, tongue in cheek.

FROM E-COMMERCE TO DISCOVERY COMMERCE

Placing orders via an app or a webshop is a well-considered and fairly rational process. Consumers check what they need, make a list and submit it. There is little or no room for impulse buys and cross-selling. Or is there? According to Facebook, the majority of online sales in the future will shift from e-commerce to discovery commerce, where the product finds the customer, rather than the other way around. Retailers and brands will make use of new creative tools to allow consumers to discover their products. Online searches will become more inspirational: instead of asking 'Show me exactly what I want', the new command will be 'Show me what will make me happy'. This requires a high degree of connectivity: friends, influencers and experts will enthuse customers and suggest what might be the best products for them. Manufacturers and retailers need to guarantee a seamless shopping experience: all the products must be immediately 'shoppable' at the moment of first discovery. Every source of friction needs to be eliminated: an integrated platform must ensure that nothing distracts or irritates the consumer, so that he/she breaks off the sale. In this way, the shopping experience, based on accumulated customer data, will become more personal, more relevant, better informed – and much more fun.

The stores of tomorrow: convenience is the new cheap

One of the possibilities being explored by many retailers is the development of new retail formulas based on more specific customer needs. All the major food retail groups now have a varied portfolio of store formats, with which they hope to capture well-defined target groups and consumption moments. While hypermarkets continue to find the going hard and supermarkets are engaged in seemingly endless price wars, local and convenience stores are booming. These are mainly city-based stores that aim at a public of busy, active and always connected professionals: people who have little time but lots of money. The bulk of their customers are made up of people living nearby, office workers from the immediate vicinity and passers-by. The stores are generally quite small but nevertheless fulfil a variety of functions: they cater to impulse purchases, emergency buys and on-the-go consumption, and in some cases even offer co-working spaces. Speed and convenience are their watchwords. Within these stores, three trends in particular are coming increasingly to the fore.

1. Bye-bye checkout!

Since the announcement of the revolutionary Amazon Go store concept in 2016 (it would take a further two years to actually get the first store open), all the major retailers have been searching for ways to achieve what has become the sector's new holy grail: checkout-free shopping. By eliminating the often irritating payment moment from the shopping process, customers can save some of their valuable time. This makes them feel good, and customers who feel good are never a bad thing. In the meantime, Amazon now has about twenty Go stores spread throughout the United States and has opened a series of similar stores in London. The technology seems to work well and the level of productivity is 'exceptionally high', according to the retail experts. Unfortunately, the development and investment costs are also exceptionally high ...

Other retailers tend to have more modest ambitions in this direction. Many supermarket chains now have self-scan apps that allow shoppers to scan their products and pay for them digitally. This technology has its supporters, but in terms of convenience it bears no comparison with Amazon Go. Albert Heijn's new 'tap-to-go' system is more promising in this respect. Customers download the app and activate their customer card or smartphone. In the store, they tap their card or phone against the price card on the shelf of the product they want, following which they can remove the product and put it in their basket. After ten minutes, the amount is automatically deducted from their bank account. After a first successful test phase at the company's headquarters in Zaandam and two stores in the Amsterdam Medical Centre, it is now planned to extend the scheme to at least three other stores in the autumn of 2021. According to AH, the technology is now 'mature'.

View Albert Heijn's 'tap-to-go' concept here:

https://youtu.be/IlUrPaWjJmE

2. The robot store

The main advantages for retailers of robot stores or automatic stores are obvious: low personnel costs and unlimited opening hours. This kind of store has a range of very different formats. Some are no more than a series of dispensing machines set alongside each other. Others are based on fully integrated robot applications. The locations are highly varied as well: from busy hotels and offices in metropolitan centres to the 'food deserts' of the thinly populated regions of the United States, Norway or China.

In Zurich, for example, Valora has opened a checkout-free convenience store in the middle of the city, offering a carefully selected range of more than a thousand products. Known as the 'avec-box', it focuses on fresh food for on the move, everyday products like coffee, and other household staples that everyone needs. Only people who have registered via the avec app are allowed to enter the store, to which access is gained by scanning the QR code at the entrance. Payment for the purchases is also made via the app.

The French Auchan chain has recently brought its unmanned store concept, Auchan Minute, from China to Europe. This is a mini-supermarket about the size of an average goods container, which is open 24 hours a day. Customers must first identify themselves via a special account. Once inside, they can choose their products and scan the barcodes into a self-scan checkout unit. The payment is made by smartphone. The autonomous store of Monoprix (the Casino Group) is called the Black Box. It measures just 18 m^2 and contains roughly 300 references, mainly snacks, household articles and toiletries. The interesting thing about this format is the fact that there is no app to download before you can start shopping. You can enter the store using just your credit card. You then take the articles you want and enter them into a touchscreen. The shop has no cameras, but it does have super-sensitive weighing scales.

Not wishing to be left behind, Albert Heijn has opened a fully digital Albert Heijn To Go. This mini-store is smaller than most, but its 14 m^2 is packed with high-tech gadgetry, which allows customers to buy products as if they were taking them out of their fridge at home. There is no scanning and no checkout. The technology comes from the Chinese software company AiFi, which has recently fitted out the world's largest automated supermarket in Shanghai. This has an area of 376 m^2 and makes extensive use of smart cameras. The company has also concluded a deal with the Dutch Wundermart chain for the installation of unmanned concepts in hotel lobbies and other similar locations.

Take a virtual tour of Monoprix's Black Box:

https://youtu.be/zLdKzJrZqqw

3. The multifunctional store

The stores of tomorrow will take over part of the function of (take-away) restaurants, by selling fresh ready-made meals and offering customers the opportunity to eat them on site. This can range from a simple breakfast of fruit juice, coffee and croissants at Carrefour Express to full meals at Jumbo City and Jumbo Foodmarkt. Under the name of Bon Appétit, Carrefour has launched a large restaurant in Paris, but also opened smaller Carrefour Cafés in its hypermarkets. Other new convenience formulas not only combine catering and retail functions, but also offline and online functions, by serving as a collection point for online orders.

Perhaps the most remarkable of these formulas is the American Choice Market, which combines almost every conceivable function at its store in Denver. It is an impulse store, fresh food store, automated store and fast-casual restaurant all rolled into one. It sells some 3,000 references, but 60% of its turnover comes from the sale of meals, either pre-packed or made to order on site. Another 30% of the turnover comes from online sales. Following a lengthy test phase, founder Mike Fogarty is now planning new openings. He has also developed an unmanned robot concept under the name of Choice Mini-Mart. Has the age of the hybrid store finally arrived? It certainly looks that way.

WILL ROBOTS REPLACE HUMAN EMPLOYEES?

In the future, it is possible that robots will not only play a role behind the scenes – for example, in distribution centres – but also on the shop floor. Various retailers have carried out tests in recent years, with increased efficiency as the most important objective. Robots can work 24/7, never get tired or sick, do exactly what is asked of them and never strike – at least until their battery runs out. Walmart, Ahold Delhaize and Carrefour are among those who have experimented with robots for different reasons. Some have been used at store entrances to give customers a friendly welcome, or to show them the way, or to give them a store plan, or to answer simple questions. Others have been used to scan shelves – the robot can signal which products are almost sold out and need to be urgently replaced – or to mop up products accidentally spilt by customers. In this way, robots can become smart assistants for their human colleagues.

Applications where robots replace human beings entirely are still seldom. But it is not for the want of trying. FamilyMart and Lawson, two Japanese chains of local supermarkets, are already using robots to fill the empty shelves in their stores. The robot in question, named Model-T, was developed by the Japanese start-up Telexistence and has arms that can reach up to two metres high. Equipped with cameras, microphones and sensors, it is capable of picking up objects of different sizes and shapes and putting them down in different places, so that the device can restock shelves with products in bottles, tins and bowls.

However, these robots also have their shortcomings. Leaving aside the way they look – they have a curiously cat-like appearance – they still need to be operated at a distance by humans, until such time as artificial intelligence teaches them to imitate human actions more perfectly. For much the same reason, it is advisable not to give the current three-fingered models soft products such as fruit or bread, since they lack the necessary delicacy of touch. Moreover, it currently takes a robot eight seconds to do what a human can do in five.

Even so, the Japanese have great faith in the potential of this technology, particularly in view of the acute shortage of labour in their country. With a population where one out of every three people is older than 65 years of age, it is not easy to find human beings who are willing to fill supermarket shelves. This means that in Japan even distance-controlled robots are an interesting option, since it makes it possible for one person to fill multiple shelves with multiple robots at the same time.

Telexistence seeks to exploit this market by renting rather than selling its robots, and this at a price that the start-up claims is equivalent to the wages of a human shelf-filler. The company also expects that in future the controllers of the robots will be located in low-wage countries, since these 'pilots' can do whatever is necessary from anywhere in the world with the aid of VR glasses, a joystick and the robots' own cameras and sensors. And thanks to the robots' microphones, it will even be possible to talk with customers and staff in the distant stores.

Watch the robots at work here:

https://youtu.be/UxWH5XAcFnM

The U-turn of the supermarkets: fewer products, more solutions

Has the physical store had its day? No, of course not! But they do need to change direction – and for the giant retail 'supertankers' this is by no means easy. The expected growth of e-commerce will certainly have an impact on supermarkets, so that in the years ahead they will need to have a radically different look. What will the most important aspects of these changes be? The number of products on offer will be drastically reduced, because online supermarket sites will offer consumers an almost endless range, so that there is no point in repeating this at the physical level. For each product category, the physical stores will have enough with a dozen or so carefully chosen stock keeping units, which will provide shoppers with quality and convenience. These same shoppers will order the rest of their needs via a touch-screen and will either pick them up later or have them delivered to their home.

As a result, the supermarkets of the future will have fewer aisles and fewer shelves. The dry goods section, which today forms the heart of the store – the so-called centre store – will need to make way for a food court with freshly prepared meals and a catering facility where they can be consumed, if so desired. It will also be necessary to have a section with new products that can be changed every two weeks, so that shoppers can be continually surprised and made happy. A farmer's market supplied by local growers can also be an important magnet for attracting people to the store.

At the same time, technology will play an ever-increasing role. A concept that involves a collection point will become standard. 3D-navigation, which will allow shoppers to link their smartphone to the shop layout, will make finding what you want ever more easy. Customers will also receive personalised recommendations on site, thanks to the use of artificial intelligence. Traditional checkouts will disappear altogether, to be replaced by self-scan tills and automatic payment. If retailers wish to remain competitive, they will need to make their processes more efficient via digitalisation. In future, anything that can be automated must be automated.

The hypermarket in search of hyperconnection

The numerous exercises carried out in recent years by Carrefour and others are typical of the modern retailers' attempts to reinvent the hypermarket. Before the breakthrough of the endless shelves of e-commerce, the hypermarkets were the kings of the mega-range. Their huge food assortment was a catalyst for non-food sales. Add to this a vertical offer of fashion goods and the result was a dramatic increase in margin per square metre. For many years, 'everything under one roof' was the order of the day and success was guaranteed. Customers were happy to make time in their busy agendas to buy the bulk of their domestic grocery needs at a single location. But not any more.

The retail landscape has changed quickly and drastically. Non-food has seen the irresistible rise of 'category killers' like MediaMarkt and Decathlon, but also the growth of specialised bookstores, toy shops and do-it-yourself chains. The subsequent emergence of e-commerce has simply speeded up this process. Result: a previously unseen decline in the number of visitors to hypermarkets. At the same time, competition in the food sector has also increased, putting margins under ever greater pressure.

Given this scenario, the food retailers held on tight to their final hope: their private labels. For some players, their house brands now represent a 60% share of sales, including fresh products. The traditional supermarkets now have a greater own brand assortment than the original specialists in this field. For this reason, Aldi and Lidl were also forced to change their approach, which they did by exploiting their extensive store network as their trump card. For these hard discounters, each store is usually assessed on the basis of a particular flow of customers. When a store becomes saturated, they open a new one somewhere nearby. So when they saw the competition starting to edge into their territory, they simply reset their marketing focus: the price champions now became the new standard-bearers of local store convenience, with bakeries and an extensive fresh food section as their new weapons. But their biggest weapon of all was the expanded availability of A-brands, which meant that consumers had no need to travel miles and miles to get their favourite pot of Nutella or bottle of Coca-Cola. By contrast, the hypermarkets are often much further away, with all the traffic problems this implies.

The store as a physical marketplace

Put in simple terms, the huge 'everything-under-one-roof' stores are now stuck with too many 'dead' square metres that they no longer need. So is this over and out for the hypermarkets? Not necessarily. And certainly not if you look at the current situation more abstractly. You can regard a hypermarket as a private shopping centre. Or as a marketplace. Or as a physical platform. There are lots of possibilities.

The model that the Extracoop chain in Italy is currently experimenting with is smart and has interesting potential. Chefs prepare meals 'live' in front of the shoppers, who in this way can learn some of the tricks of the culinary trade and then eat the resultant meals on site. There are a number of these so-called 'Good Pause' eateries in different formats, including a pizzeria, a restaurant, an ice-cream salon and a bar. Healthy fast food and convenience are rapidly gaining ground in the modern-day retail environment and these are the aspects that this concept tries to exploit.

In terms of its non-food component, Extracoop has a large wellness section that devotes considerable attention to health, nutrition and personal care. The parapharmacy contains more than a thousand different products from leading brands in the world of cosmetics and health food. People seeking professional advice can receive it from the pharmacists who are always present. Its optical section is also a real eye-opener, offering free eye tests and a huge range of glasses, contact lenses and related products.

To extend customer choice even further, other shops selling other products adjoin the entrance to the supermarket. Where there is a margin to be gained, Coop manages these shops itself. In this way, it also avoids unsightly empty premises for its more successful surrounding lessees. Supermarkets like Extracoop are starting to realise that it is becoming increasingly less feasible and less desirable to meet people's entire shopping needs in a one-stop store. In this respect, it is also worth noting that there are no separate checkouts in the subsidiary shops: payment is made in the hypermarket. This makes it possible for the shop staff to concentrate on their main task: serving the customer.

This emphasis on service also finds expression in the wide-ranging service package from which Extracoop members can benefit. For example, subscribers can gain access to Coop holiday deals at super-competitive prices, while others might prefer to order a show cooking demonstration in their home. By pampering its customers in this way, Coop hopes to turn them into brand ambassadors. With a typical touch of Italian romance, the company refers to this as 'the distribution of emotions'.

The various aspects of the concept that are designed to give the supermarket an impression of 'specialisation' are probably still too optional and are therefore unlikely to make the difference in terms of binding customers. What is innovative, however, are the subsidiary margin-makers, such as the optician's shop and the beauty salon, which from the outside appear to be specialist stores in their own right, but are actually run by Coop. As a result, the sales personnel can concentrate fully on guiding the customer to a purchase, assisted by inline retail in the form of kiosks in the store that give access to an endless assortment.

In theory, the same thing should be feasible for food products. So why not have subsidiary shops with local butchers, regional products or organic farmers? Physically, or with the help of digital tools, it is possible to give stores like Extracoop a human face. This unique mix of technology with a human touch can help to revive the fortunes of the hypermarkets and turn them once again into places where people meet.

WHAT CAN FOOD RETAILERS DO WITH EXCESS SHOP SPACE?

As a result of the growth of e-commerce and the convenience channel, many traditional hypermarkets and supermarkets are confronted by the fact that they now have too many square metres. What options do they have to make use of this excess space profitably?

- Reduce the size of the store and lease out the free space to other brands. This not only provides rent income but also has the potential to increase traffic in your store. The lessees can come from many different sectors. Perhaps a hairdresser's or maybe a sushi bar? In a number of its French hypermarkets, Carrefour outsources its electrical goods department to FnacDarty. In Belgium, Makro works in the same way with MediaMarkt.

- Part of the store can be converted into a dark store for preparing online orders for home delivery or click-and-collect.

- Develop your own new activities that can complement your regular product offer: a restaurant, a demonstration space, an event room, etc.

Carrefour's search for a new identity

Carrefour is testing out a number of these possibilities in different places. Their main conclusion is that you do not need to pay for all these excess square metres yourself. In Belgium, Makro has reached the same conclusion and now rents out its unwanted space to MediaMarkt, a hip electrical brand that brings more younger consumers into the store. In other words, win-win: you reduce size, increase profitability and get a category killer in-house that pays you rent! Carrefour does something similar in its domestic French market, renting out it electrical goods department to the specialist Darty brand. In its Belgian and Italian hypermarkets, Carrefour offers a corner to Decathlon, while in return the sports retailer is testing out a Carrefour organic section in one of its stores.

One of Carrefour's largest 'foreign' hypermarkets is at Finestrat in Spain. Here the group is testing out the concept of a 'mercato', a kind of in-store street market with a bakery, sushi bar, pizza and hamburger stands, and a beer tent. The concept still needs further development and at the moment comes nowhere near matching, for example, the Mercato del Erbe in Bologna, where you can not only smell the herbs as you approach but also sense the authenticity of the other traditional craft products that are on sale. Even so, there is nothing wrong with Carrefour's basic idea. Instead of trying to do everything yourself, collaborating with local traders and farmers is a smart step in the right direction. The original is usually better than the copy. In Belgium, Delhaize has understood this as well: the supermarket chain now works with health food specialist Foodmaker, which attracts younger customers like a magnet.

Carrefour also realises that the days when one concept could meet everyone's needs are over. Stores now need to adjust themselves to the requirements of their market area. In regions where purchasing power is an issue, the retailer is now experimenting with a new concept called Essentiel, which places an emphasis on basic products at very competitive prices. In other words, a kind of discount hypermarket. In other stores, the chain has created an outlet department. Where the public is more sophisticated and has more money, Carrefour focuses extra attention on its organic section and a wider range of top quality ready-made meals. It also offers space to local growers and entrepreneurs. In general, the aim is that the group's hypermarkets should become less 'hyper' and more 'market'.

Inevitably nowadays, the physical marketplace concept, with its shops-in-shops and specialist corners, also has a digital dimension. In 2020, Carrefour launched its own online 'food marketplace'. The intention behind the project is to strengthen the range on the Carrefour.fr website by supplementing it with the products of other suppliers that reflect popular food trends: organic and natural products, products that are good for your health, gastronomic products, local or regional products, world products, gluten-free and salt-free products, etc. The basic product categories are food (domestic groceries), drink (wine, beer, spirits and soft drinks), hygiene and beauty products, pet products and child care products.

Coop's 'smart' supermarket in Milan

In Milan, the Italian Coop chain has created a high-tech supermarket that makes use of tools like augmented reality to better inform shoppers about their eating and purchasing behaviour. The ultimate aim: sustainable consumption.

The store covers 1,000 m^2 and sells some 6,000 references. Part of its offer is displayed on intelligent display tables. When you move your hand in the direction of a product, a screen above the table automatically gives you information relating to that product: where it comes from, its nutritional value, if it contains allergens, how to dispose of the packaging, etc. This is made possible by the use of Microsoft Kinect sensors, which pick up the shopper's movement.

The store also has a number of interactive totems with touchscreens. Another eye-catcher is a 20-metre long video wall with 54 monitors, which provide shoppers with real-time information about the products in the store, the promotions on offer, sales statistics, recipe ideas, etc., as well as displaying posts on social media about the store.

This futuristic supermarket was designed by Carlo Ratti, a professor at the Massachusetts Institute of Technology (MIT): 'Every product tells a story,' he explains. 'In years to come, it will be possible to know everything about the apple you are looking at: which tree it was grown on, how much CO_2 it produced, the chemical treatment to which it has been subjected, and its journey to the supermarket.'

He expects that data of this kind will lead to better informed and more sustainable patterns of consumption. A fully transparent supply chain must make shoppers more aware of the limits of our natural resources. In turn, this can stimulate the purchase of fresh, local products.

In spite of the dominant presence of digital applications, the Milan supermarket looks anything but 'cold' and 'technical': the store layout and decoration make use of low fixtures and warm materials, such as wood and ceramic tiles, with the aim of producing a pleasant but also an efficient shopping experience.

Scan the code for more info about Coop Italia's vision for the supermarket of the future:

https://youtu.be/BhNEGu-1VpA

Experience squared: FICO Eataly World

Forget Disneyland. In Bologna, there is a theme park designed especially for foodies! FICO Eataly World is a mecca for all lovers of Italian food. In 2017 and 2019, RetailDetail visited the park with managers from the food sector during an inspiration trip. In Italian, FICO literally means 'fig', but in this case it is also an abbreviation for *Fabbrica Italiana Contadina*, which can be translated as 'the association of Italian agro-food partners'. The park covers 10 hectares and was created by the Italian Coop supermarket group in collaboration with Eataly, a distributor of typically Italian products. It is the largest agro-food park in the world, where (amongst many other things) visitors can learn how to make pasta or see how dogs are trained to find truffles.

In fact, it is possible to see every aspect of the food production chain in fascinating detail, with the aim of inspiring and educating people, whilst at the same time giving them a fantastic experience. The park's 20,000 square metres of agricultural land and stables, its mini-factories and its interactive workshops allow visitors to explore every aspect of Italian biodiversity and cuisine. Of course, it is also possible to sample all these Italian goodies in one of the 45 restaurants and *trattorias* or in the food market.

The founder of Eataly is Oscar Farinetti. Since 2007, he has been a pioneer in the integration of food retail and food service. The park is the apotheosis of his philosophy. Its giant stores are a cross between a *mercado*, a food hall and a supermarket. They are places where visitors can eat, buy products or even learn how to cook (with, of course, pasta and pasta sauces as the 'classics'). The fresh products that are not used on any given day are used to make take-away meals for the following day or are used in the park's other eateries.

The concept seamlessly matches the transition currently taking place in the retail sector, where the consumer's experience is now more important than the products he buys and increasing attention is being devoted to the origin of our food, with the supermarkets in turn responding by becoming more like caterers or even restaurateurs. It sounds great, but is FICO Eataly World financially viable? Its 6 million visitors each year spend on average between 20 and 25 euros per person, allowing the park to just about break even.

Bridge: the food laboratory of Migros in Zurich

A mix of supermarket, gastronomic hotspot and event location: that is what Bridge wants to be. This new store concept was launched by the Swiss food retailer Migros at the start of April 2021 in the centre of Zurich. Migros has thought carefully about the name. Bridge aims to be precisely that: a bridge between people and food. And this is no coincidence: Gottlieb Duttweiler, the founder of Migros, also sees himself as a builder of bridges between producers and consumers.

'In the current market environment we are confronted with the question of what the bridges of the future might look like. Our vision for this location is to use our own strengths to unite a wide-ranging offer of products and services in a single location in a manner that creates new platforms,' says Jörg Blunschi, Managing Director of Migros Zurich. Put simply, Bridge is an urban meeting place where you can enjoy time with friends, buy fresh and seasonal food, and enjoy exciting culinary experiences.

The store has an area of 2,000 square metres spread over two floors, where visitors can find a surprising and constantly changing assortment of seasonal, local, traditional and organic products, from sustainable coffee and hand-made soap to vegetable desserts. There are pop-ups run by a varied range of food start-ups and street food stands where your meal is prepared in front of you while you wait. In the bistro a new chef takes over every three months. Shoppers can also take part in wine-tasting sessions and workshops on diverse themes. There is even a function space that can be hired for customer events, birthday parties, private occasions, creative team buildings, etc.

The entire complex is fitted out in warm, sustainable and natural materials, while the store concept is modular, which allows its furnishings to be adjusted to suit the needs of specific situations. Technology does not impinge visually, but plays a major role in supporting Bridge's story. In short: well worth a visit!

Why the supermarket of the future will also be a data centre

Just imagine it: a store that recognises a customer the moment that he or she walks through the door. Better still: a store that can tell from your behaviour out on the street outside that you are planning to come in. No, this is not science fiction. Recognition software for vehicles is already in use and facial recognition technology is becoming increasingly reliable (although the privacy implications remain an issue). In future, it is perfectly conceivable that supermarkets will recognise frequently returning customers and then link in with their known preferences and past purchasing history. Has the customer come to do a weekly shop or is it more likely to be an impulse or emergency buy? The store will also be able to establish this too, based on the customer's route and interaction with the different products.

Taking things another stage further, technology already exists to identify the emotional mood of customers, based on their movements and facial expressions. This offers the possibility to suggest particular products that match those moods (via their smartphone, the public address system, etc.) or to intervene when a negative experience threatens to occur (long queues at the checkout, difficulty in finding a particular product, panic when searching for a lost child, etc.).

In this way, the individual shopper's experience of the store can be personalised. The interactive screen at the cheese counter will 'know' that the customer likes strong-tasting cheese. The delicatessen section will 'remember' that he/she likes smoked chicken. The floristry corner will 'remind' you that tomorrow is your wedding anniversary. And so on. At the end of your shopping, the store's knowledge of your financial details will provide you with a pain-free payment experience.

This is the image of the future sketched by *The Store of the Future* report produced by the Coca-Coca Retailing Research Council (CCRRC), a research body sponsored by the soft drinks giant to carry out studies on behalf of its member retailers.

According to the report, physical stores will continue to play an important role for all the current stakeholders (customers, retailers and suppliers), but it will be necessary to achieve a more seamless interaction with the first of those three actors. In practical terms, this means putting the customer in a position that he/she can easily and fully satisfy his/her needs at all times and in all places: when visiting the store or ordering online, whether at home, at work or on the move. The retailer of the future cannot and must not attempt to separate the store from the digital world.

In short, supermarkets must become intelligent, which means they will need to specialise in the collection and analysis of data. This will have consequences for store design, the training of staff, and collaboration with suppliers. The fascinating CCRRC report offers a detailed examination of the implications for aspects such as logistics, automation, robotisation, store layout, marketing and personnel policy.

The report also emphasises that the future is not just about technology. One of the biggest retail risks of the 21st century is that the shopping environment will become predictable or even boring. Many stores are designed with maximum efficiency in mind: to fill the shelves quickly, to prevent theft, to eliminate other potential problems, etc. Very few of them devote sufficient attention to concepts like inspiration, convenience, speed of navigation, customer surprise or even customer entertainment.

The purpose of modern supermarkets is to make routine shopping as efficient as possible. But it is precisely this aspect that may prove fatal to them in future. This risk is most acute in what we generally refer to as the centre store, the section in the middle where dry food products, drink products, household products and care products are all concentrated. These are the products that are most under pressure as a result of changing consumer preferences, generated in part by increased online purchasing. The supermarkets of tomorrow will need to be more flexible, so that they can respond to these changing expectations.

Has Amazon cracked the supermarket code?

The retail sector is currently undergoing the 'amazonising' of the value chain. Amazon has succeeded in creating a gigantic, seamlessly integrated and worldwide supply chain with a wide range of delivery options. This success is based primarily on its immense wealth of data, which is used to continually optimise the chain by further refining its offer and even predicting future purchases, so that its factories and warehouses can respond to the needs of the market perfectly. It is impossible to overestimate the impact of this model on other sectors.

That being said, the food sector is proving a hard nut to crack, even for a retail giant like Amazon. The food market is of huge economic and strategic importance for the organisation: the daily and weekly purchase of food, drink and other household products is not something you can ignore if your ambition is to be 'the everything store'. But to achieve this ambition you need to have a successful physical store concept – and that is the problem. When, after much delay and countless PR announcements, Amazon finally opened its first checkout-free store with 'just-walk-out' technology in 2018, some observers thought that it was a game-changer. The concept is certainly revolutionary. To enter the store, customers only need to scan a code that they can find in the Amazon app. Once they have finished shopping, they just walk out again: there is no checkout and payment is fully automatic. Since the launch of this first Go store, Amazon has further refined the technology, which makes use of hundreds of cameras and sensors, as well as artificial intelligence. Yet for all its innovative brilliance, the commercial impact of this fully automated concept has been limited. By the spring of 2021, there were only 25 such stores in operation.

Undeterred, Amazon continues to experiment with new formats. In the summer of 2020, the first Amazon Fresh supermarket opened its door in Woodland Hills, a suburb of Los Angeles. This store combines a very traditional assortment of products with innovative technology, including a smart shopping trolley and (again) a cashless checkout. Amazon has big plans for a new chain of physical supermarkets, in addition to its high-end Whole Foods stores (acquired in 2017) and its Amazon Go convenience outlets. The new formula wants to offer competitive prices for a range of products that includes well-known A-brands, Whole Foods 365 house brands, the Fresh house brands and a wide variety of other fresh products. A ready-made counter will prepare fresh meals, toasted sandwiches and rolls, pizzas, roast chicken and fresh bread, made while you wait or ordered in advance online.

Shoppers in these large stores (1,000 m^2) will make use of intelligent Dash trolleys, fitted with sensors and image-recognition software. This means that the trolley can identify the products the shopper puts into it and add up their total cost. Queuing at the checkout is no longer necessary and payment is automatic. Amazon Prime members can order products online from the store for delivery or collection later the same day. There is also a collection and return point for online orders made via amazon.com. Alexa kiosks are on hand to provide customers with any information they need. All they need to do is ask 'Alexa, where can I find the ketchup?' and they will immediately be given an answer. Shoppers at home can also have Alexa draw up a shopping list for them.

In the meantime, Amazon has exported a smaller version of its Fresh concept to London. It is the retailer's first automatic store outside of its domestic market. The smart shopping trolleys are not yet available but all the other features are there. This heavy investment makes clear that Amazon still wants to conquer the food market but now understands that this will not be possible without a sizeable physical footprint. In other words, it needs 'real' supermarkets and a significant store network. For this reason, observers expect Amazon to be active in the takeover market in the years ahead, both in the US and in the highly fragmented European market, where the retail giant might play a leading role in a new wave of consolidation.

Discover here what Amazon Fresh London looks like:

https://youtu.be/ETh4HdBCRKk

How China is reinventing the supermarket

But if we want an idea about what the supermarket of the future will really look like, perhaps we need to turn our attention to China. In terms of innovation, Alibaba's Freshippo supermarket chain (previously Hema Fresh) is streets ahead of its Western rivals, with its perfect mix of online, offline, technology and human experience. The chain has five different formats, which vary from automated take-away restaurants and fresh markets to giant shopping centres. Alibaba makes a clear distinction between its different locations and their different functions, with a hierarchy of cities (ranging from large and prosperous to small and backward) and adjusted ranges for residential areas and business districts.

One of the large centres is the Freshippo Mall in the city of Shenzhen: a complex measuring 40,000 m² over three floors, where in addition to buying food customers can also find entertainment, children's activities and catering facilities. Moreover, all the products on offer can be delivered to the customers' homes (if they are within a radius of three kilometres) in just an hour. The underlying idea is that Freshippo should become both a traditional one-stop shop and a community builder.

In more residential areas, the most popular concept is the Freshippo Farmers' Market, where Alibaba attempts to give the traditional covered markets of the past a new look. The fresh products are still there, but so too are digital checkouts and home delivery within 30 minutes. In inner city centres where the chain currently has no supermarket, use is made of the Freshippo Station, a fulfilment hub that primarily focuses on the home delivery of online orders of fresh food within a radius of a kilometre and a half.

More in keeping with the traditional supermarket concept is the Freshippo Mini. These stores have an average size of between 300 and 500 m^2, making them ten times smaller than the standard Freshippo supermarket. Alibaba readily admits that the Minis tend to be concentrated in remoter and less prosperous cities, where a wide range of high quality goods is less needed and more difficult to supply. The chain hopes to use these convenience stores to attract independent entrepreneurs who currently run hyper-local grocery businesses.

Because blurring is also becoming a trend in China, more and more of the Freshippo supermarkets are opening high-tech 'on-the-go' restaurants. Freshippo F2 – standing for 'Fast and Freshly-made' – offers lunches and snacks in a buffet format to city centre office workers, while Freshippo Pick 'n Go dispensers offers other city goers 'breakfast out of the wall' all day long.

Notwithstanding their many differences, these concepts all have one thing in common: they enshrine the principle of 'new retail', Alibaba's conviction that digitalisation must also be integrated at all levels into physical retail. E-commerce cannot work without physical stores, but the reverse is also true. This is something that the Chinese understand better than anyone else. As a result, Freshippo consistently applies a resolute 'mobile first' approach, in which self-scanning and unmanned checkouts are standard.

Mobile ordering for collection or home delivery is a fundamental part of each store's function, whatever its size. Every supermarket is a local fulfilment centre, where staff move between the shoppers to collect items from the shelves for the conveyor belts, so that couriers can deliver them to people's homes, often within half an hour. As for the physical shoppers, they can scan in their products and then have them delivered by an ingenious robotised tube system in the roof to the exit or the restaurant kitchen, so that they can take them home, have them delivered or have them cooked on the spot to eat! This is the ultimate in convenience.

Alibaba's great rival JD.com has also developed a high-tech supermarket. 7Fresh integrates e-commerce technology into the shop floor environment and aims to have a thousand stores throughout China. The first 7Fresh was opened at the end of 2017 in the Dazu Plaza Shopping Centre in Peking, close to JD.com's headquarters. The supermarket is some 4,000 m² in size and derives 75% of its turnover from the sale of fresh products, including many imported goods that Chinese consumers regard as exclusive: Spanish Iberico ham, fresh shellfish from Japan, French patisserie ... Customers who so wish can have their purchases converted into meals that they can eat on site in the restaurant, so that the JD.com concept once again combines retail with food service.

To adjust its assortment to reflect changing consumer expectations, 7Fresh makes use of big data analytics. This advanced technology makes it possible for the retailer to give its customers a personal and educational shopping experience. 'Magic mirrors' automatically display information about the origin and nutritional value of the products that people remove from the shelves, in a manner that is similar to the technology introduced by Coop in Milan back in 2016.

Scanning takes place using a mobile app and digital payment technology settles the bill. 7Fresh also arranges home delivery of orders within a radius of five kilometres around its stores, often within 30 minutes. In this way, the 7Fresh concept, according to JD.com, illustrates how traditional retailing is nowadays driven by e-commerce technology. It is also worth noting, says the company, that the e-commerce boom in China owes much to the lack of good and well-stocked physical alternatives, a shortfall that JD.com is now doing its best to make good. The retailer proudly claims to have the most efficient purchasing, storage and logistics operations in China, with a guaranteed cold chain to the last mile. At the same time, JD.com is also working with a growing number of partners to use blockchain technology to trace and record every step in the supply chain, from production to delivery. Interestingly (and importantly), one of the shareholders in JD.com is the American food giant Walmart.

Since the opening of its first store, JD.com has modified its concept on more than one occasion. The newest spin-off is 7Fresh Life. These smaller stores (300 to 400 m²) are open 24/7 and combine a quick-service restaurant, a fresh food outlet and a convenience store.

These are all fine examples of 'new retail' in action and demonstrate how the shopping experience in China is being digitalised – and fast. Indeed, perhaps that speed is the most remarkable aspect of China's extraordinary surge of retail innovation. But are these concepts a blueprint for the rest of the retail world, as Alibaba claims? Probably not, or at least not in full. There are some elements that will be difficult to copy in the more mature Western market. One thing, however, is certain: the inspirational examples of Alibaba and JD.com are certainly giving retailers in the US and Europe a lot to think about!

Discover Alibaba's different Freshippo formats here:

https://youtu.be/xWZBAx6sOX4

WHAT NEW RETAIL MEANS

The most important innovation introduced by Alibaba in its fresh food supermarkets is the integration of the online and offline channels on the basis of big data, mobile payment and smart logistics.

1. Physical shopping is now done with a smartphone in your hand. This allows you to call up product information, receive recipe ideas, make reservations and place orders.
2. The convergence of different retail formats makes it possible to benefit from different experiences at the same supermarket location. You buy raw ingredients to prepare yourself, you buy (partly) prepared meal components or complete meals to heat up at home, or you have the ingredients prepared for you by chefs, and you reserve a place in the in-store restaurant. Convenience doesn't come any easier than this.
3. The supermarket also serves as a distribution centre for home deliveries within a radius of five kilometres in just half an hour.

Five takeaways: the supermarket of the future must become more of a market

In the previous chapter, we examined various options and initiatives that all share a common purpose: to make it possible for the food retailer to better serve the consumers of tomorrow. The rapid rise and growth of new digital distribution platforms and channels threatens the market dominance of the traditional players, irrespective of their size. To remain relevant, these players will need to become platforms themselves. These evolutions will force the supermarkets to reinvent themselves, and it will need to be a radical process. The following five takeaways show the direction that the supermarkets of the future will need to take.

1. Optimal use of the available square metres

After more than 50 years of little or no change, the supermarkets of tomorrow will look very different. Food retailers will need to make the best possible use of the square metres of space available to them. This means less space for the display of products – because such purchases will largely be made online or even automated – and more space for solutions, inspiration, entertainment and logistics. This will require a different kind of store layout and design, in which products are no longer displayed by type, but in function of their use; no longer in function of logistic requirements, but in function of consumer expectations and/or the moment of sale. Why is fresh pasta currently not found alongside dry pasta? Or fresh vegetables alongside frozen ones? For consumers, that would make more sense. Moreover, the store layouts of the future will no longer be fixed but variable. This in turn will have an impact on how stores are built: Stamhuis, the store-builder for Albert Heijn, has even developed a method that allows stores to be dismantled and reassembled in the record time of just one week, thanks to the extensive digitalisation of the entire preparation process

2. Online and offline merge into each other

New technology will appear in the supermarkets of the future in response to new methods of shopping. The smartphone will offer access to all kinds of information about products: ingredients, nutritional value, allergens, raw materials, processing, transport, etc. It will even supply recipe ideas! However, other advanced data applications will also be necessary if supermarkets want to function properly in the years to come: personalised promotions, dynamic price setting, temporary flash sales and intelligent cross-selling will become the new normal. Instead of taking products from the shelves and putting them into their trolleys, customers will scan the products and pay for them automatically, following which the products will be waiting for them at the exit or will be delivered to their homes.

Shoppers will have maximum purchasing choice: they can order products online from their own front room and have them delivered to their own front door; or they can order part of what they want online and shop for the rest in a physical store; or they can buy everything in a physical store and take it home or have it delivered. For example, you might want to order your dry products online and have them delivered, but you may prefer to pop into the store for your fresh food or to pick up some recipe ideas. Or have you run out of something that you need quickly and find it easier to call into a store on your way home? Everything is possible. Online and offline have become one.

3. The dark store in the spotlight

To make this possible, part of the available store space will need to be converted into a dark store: the logistical hub where online orders are made ready in record time, both for customers who come to collect them at an agreed time (click & collect) and for people living nearby who have asked for home delivery. Although it is called a dark store, it is a good idea to shed some light on it: insert a glass partition wall, so that people in the supermarket can see what is happening. It is all a question of transparency...

4. Back like it's never been away: market 4.0

The retailer needs to make use of the remaining store space in function of its ambitions and positioning. A 'live cooking' in-store restaurant that has dishes on its menu made from fresh ingredients available in the store strengthens the 'fresh' image and expertise level of your operations. A large and varied fresh market, packed with produce from local suppliers, will strengthen your ties with the local community. Cooking lessons, demonstrations, tastings and workshops will work as a magnet for food-minded customers and increase their loyalty. Modular store layouts make it possible to constantly vary your offer, with a frequent rotation of themed actions that will surprise, please and entertain your customers. Put simply, the market will be put back in supermarkets (although it never really left). But this time, it will be in a digitalised form: the retail of tomorrow is a platform.

5. Taking sustainability to the next level

It goes without saying that the supermarkets of the future will need to set new standards of sustainability. Circular and passive building design, energy neutrality, ecological materials, ecological cooling, water recuperation, and the intelligent monitoring of all parameters will become commonplace. More than that, they will also be a prerequisite for success. The assortment will need to be built around greater attention to fair trade, organic products and short chains, with full transparency about suppliers. Future-proof supermarkets must avoid unnecessary packaging and take measures to stop food loss by processing and discounting surpluses intelligently in relation to their 'use by' date and by entering into open partnerships with charitable organisations. Only in this way can the food retailers of today be ready to face the challenges of tomorrow.

In 2022, a prototype of what the supermarket of the future might look will be opened in France. In this video Accenture and Le Projet Hope give us a first glimpse:

A22

https://youtu.be/MFg3rBGVb_g

Europe, sandwiched between America and China

Europe is not exactly top of the class when it comes to innovation in retail. On the contrary: the two other retail superpowers are miles ahead and see Europe increasingly as easy prey. Does Europe risk being sandwiched between American retail-tech on the one hand and Chinese 'new retail' on the other?

The greatest innovator in the world...

Whenever the word 'innovation' is mentioned, people automatically think of Amazon. To support its ambition to become 'the everything store', Amazon spends 20 billion dollars a year on research and development, which is some 10% of its global turnover. In 2018, this made it the most innovative company in the world, ahead of Alphabet (the mother company of Google) and the pharma-giant Johnson & Johnson. In part, this investment has found expression in a new and radical omnichannel food strategy. What started in 2017 with the acquisition of Whole Foods and the first modest delivery service with Amazon Pantry, has since been transformed into a network of highly automated (and preferably checkout-free and personnel-free) convenience outlets in the shape of Amazon Go stores and collaboration with other major retail players, like Casino and Morrisons in Europe.

The key question is whether or not the retail giant from Seattle has finally found the concept for the supermarket of the future with its newest venture: Amazon Fresh. This hybrid store certainly combines all the most important contemporary retail trends. A large part of the floor space is devoted to the processing of online orders, but without compromising on food service and customer experience. Pizzas and fresh bread are made on site, and can be ordered in advance via the store app. Smart shopping trolleys immediately identify what customers have bought and automatically settle the bill with no need to pass through a checkout.

... is banking on the fridge

According to the authors Berg and Knights (2019), Amazon will indeed succeed in its ambition of breaking through into food by 2025. The Seattle juggernaut will devote all its massive resources to this task, not simply because it is the last fortress to be conquered, but also because food is the key to all other purchases. Fresh food and groceries are the most repeated of all repeat purchases, so if you can create consumer confidence in this sector you have a foot in the door of every family in the world. Love, as the old saying tells us, goes through the stomach and it is indeed true that people give their confidence to those who are able to satisfy their most essential basic need: the need for fresh and high-quality food.

In other words, for Amazon food is the key to becoming an all-embracing ecosystem surrounding the consumer. In this process, data is both the means and the end. Jeff Bezos has been dreaming for years about being able to satisfy people's grocery needs before they even know what they need themselves. Bezos will know it for them. Thanks to a series of connected household devices – from smart fridges to talking cookers – supported, of course, by the omniscient smartphone, your toilet paper will be automatically replaced before it runs out and your new supply of cola (or beer) will be delivered before you have drunk the last bottle. This concept of automatic replenishment is already a reality for Philips electric toothbrushes, Brita water filters and ink cartridges. Voice commerce – talking to your smartphone or a smart speaker in your home – forms the bridge between a fully automated system and the often cumbersome online ordering process.

Don't call food-tech a supermarket

Amazon's competitors are starting to sing the same tune. Albert Heijn, a subsidiary of the Ahold Delhaize group, now calls itself a food-tech company, while Walmart paid far too much for Jet.com simply to get its hands on the company's engineers. They were needed to give the mother ship – which in Walmart's case is a supertanker – a digital facelift. And with success: America's leading retailer is now much better equipped to take on challenger Amazon in their bitter fight for food supremacy. In 2019, Walmart's 31% of market share made it the biggest player in the online groceries sector in the US, but Amazon was close on its heels with a 29.6% market share. No other retailer scored more than 5.9% (Melton, 2020).

With only a slight degree of exaggeration it is possible to argue that the corona pandemic in the spring of 2020 came as a godsend for Walmart. When the number of online orders suddenly doubled, the online giant reacted quickly, not only by increasing its number of collection slots by one-third, but also by switching permanently to increased fulfilment and delivery from its store network. This was the omnichannel breakthrough for which the company had been preparing for years. The world's largest food retailer has also been experimenting for longer than most with self-driving delivery vehicles, both for the last mile to the consumer and for the journey between warehouse and store/collection point. But Walmart's highest expectations are reserved for its subscriber service, Walmart+, which is its answer to Amazon Prime.

WALMART INTELLIGENT RETAIL LAB

Walmart uses some of its existing hypermarkets as live test laboratories. The first of these, the Walmart Intelligent Retail Lab (IRL) in New York, is a successful store with 30,000 articles, more than a hundred personnel and roughly a million transactions each year. Customers continue to shop there every day, while Walmart continues to develop and test out its new technology on them.

These tests focus on smart image recognition (computer vision) and a limited number of core capabilities based on artificial intelligence, because the research team has discovered that many of the (more than 200 identified) use cases for AI all employ the same basic functions, such as product recognition. Roughly 85 to 90% of the team's time is therefore spent on developing scalable and cost-effective solutions, which cannot only be adjusted to suit different store formats, but also different use situations. Technologies that seem to be promising will eventually be rolled out to the rest of the Walmart chain.

Two possible AI applications on which the IRL in New York is concentrating are the reduction of theft and a reduction in the number of 'sold out' articles on its shelves, both of which can be assisted by computer vision and product recognition. More than 3,000 cameras have been installed above the store's shopping aisles. These smart cameras can recognise when someone fails to scan or incorrectly scans an item at the self-scan checkout. At the same time, possible 'out-of-stocks' are continually monitored.

Innovation is also about making choices. In New York, Walmart had to empty about 35% of its available space on one side of the building to create a data centre. This centre is equipped with hundreds of servers, whereas a normal Walmart store only has two. Simply to install all the necessary cabling, it was necessary to sacrifice 30% of the total number of stock keeping units. But it seems to have been worth it: sales have increased by 20% per annum and customer satisfaction levels have never been so high.

The decision to concentrate on product recognition in the New York tests was a deliberate one that builds on work carried out in the Walmart lab-hypermarket in the Chinese city of Shenzhen, where Walmart has developed data insight to such a level that it can accurately identify potential shoppers, can predict what they want, and can offer them appropriate personalised suggestions.

Under the motto 'A line of code can change the way the world shops', the brains behind Walmart's plans for the future are located not, as you might expect, in the US but in India. A team of thousands of software engineers, data scientists and service professionals, spread over three cities, is working to develop worldwide technological solutions, from software that can calculate the best route for order picking and the design of an e-commerce platform for Walmart Mexico, right through to ground-breaking applications for the internet of things. Are there problems with the cold chain in the US? The team in Bangalore can respond immediately. This is global connectivity at its best. Perhaps it also explains why in April 2020, right in the middle of the corona crisis, Walmart Global Tech India decided to take on an additional 2,800 personnel.

Scan here to take a look at the Walmart Intelligent Retail Lab (IRL) in New York:

https://youtu.be/sfMGtst9X1k

New retail in China

Sounds impressive? Perhaps, but there is still a long way to go. According to innovation expert Peter Hinssen, Silicon Valley is a museum in comparison with China. And Europe is even further behind. Hinssen condemns the defensive attitude of the Old Continent. 'A new cold war is approaching. And it will force us to make choices. We have held on to the Western bloc for 70 years. So how can we look to the East now? There is a revolution going on.' The language is provocative, but the message is clear: 'I'm afraid that we will need to make a much greater effort' (Van Rompaey, 2019).

Nowhere is digitalisation proceeding at such a pace as it is in China. Thierry Garnier, the former CEO of Carrefour China and now the top man at Kingfisher, used to have two pieces of fruit delivered to his office each day. Fresh, ready to eat and free of charge. It is an anecdote that speaks volumes about the retail landscape in China's major cities. Because last mile delivery in China costs just one dollar, in comparison with twelve dollars in the US, the economic logic of e-commerce is radically different. The 1.4 billion Chinese receive on average one and a half packages per week and in many of the more remote locations in this vast country people are more familiar with e-commerce than with what we in the West traditionally understand as retail.

Against this background, a new and uniquely Chinese variant of omnichannel food retail developed: Alibaba – known for its online platforms AliExpress and Tmall – refers to this variant as 'new retail'. Competitors prefer the term O2O retail, standing for 'online to offline'. This gets to the heart of the matter: the path to sales begins resolutely online, preferably via the smartphone, a device on which 95% of Chinese men and women spend an average of three hours a day. At Freshippo, Alibaba's premium supermarket concept, some 60% of its daily turnover comes from online sales.

The store as a cloud

The next step in the process is the integration of the physical store, in the first instance by using them as centres to dispatch the online deliveries to customers living within a radius of one to three kilometres. This, it seems, is the most efficient method for many categories with a high turnover rate. For example, each Walmart store in China prepares roughly 400 orders per day, not only for itself but also for JD.com, its e-commerce partner. For some items it is simply more viable for the Chinese online giant to use Walmart hypermarkets as local hubs than to store these items in its own fulfilment centres.

So what does Walmart get out of it? Fresh online orders are now its most important focal point in China. These orders are delivered to people's homes within the hour. The average time is forty minutes. Preparing an order takes ten minutes and waiting for a courier to collect it takes another ten. However, the delivery is not free. Walmart charges a fee for each delivery, primarily to make clear to the customer that this is an extra service that can be avoided if the customer makes the effort to come into the store.

How does the American retailer manage this? Through the use of cloud stores. These are dark stores that are not open to the public but serve exclusively as fulfilment centres in urban regions where Walmart has no large physical stores. Thanks to their low cost structure, these cloud stores can operate at break-even after as little as nine months. The last mile delivery takes place via a crowdsource platform that Walmart has set up, which has 5 million 'independent' couriers on its books, in a manner similar to Uber Eats. What these couriers earn is dependent on demand: at peak moments the fee is increased, so that more of them will jump on their motorcycles; when things are calm, the fee is lower.

GROUP PURCHASING WITH NEIGHBOURHOOD DELIVERY

Local (mini-)warehouses are also important for the highly popular Chinese phenomenon of group purchasing. According to Goldman Sachs, almost half of all grocery purchases in the country will be made online by 2025, with group purchases forming the majority. Although this concept first emerged as long ago as 2006 in the province of Guangzhou (Treadgold & Reynolds, 2020), it only made its major breakthrough during the recent corona pandemic.

Letting your neighbour do your shopping

To keep prices low and to make the last mile as efficient as possible, consumers order food products in groups via WeChat, the app that the Chinese use for almost everything, from sending text messages to paying for parking fines. Platforms for group purchases (tuángòu) have their own mini-platforms, on which customers can communicate their grocery needs to the person who has agreed to act as the coordinator for the local neighbourhood. These group coordinators might have their own small local shops, but often they are public-minded citizens who are prepared to receive communal deliveries and distribute them to the neighbours in their street, block of flats, etc. For this, they are paid a commission of between 5% and 10% of the cost of the bulk order.

This model sidesteps one of the main obstacles for e-commerce in fresh food products, which in China are typically bought at local markets. Outside the major cities in particular, the delivery of perishable goods is complicated and expensive. It requires heavy investment, not only in logistics but also in marketing, to generate the necessary degree of customer confidence in the system. By using local intermediaries as points of contact, these local initiatives kill two birds with one stone.

The group coordinators have a dual role. They are much more than just glorified delivery couriers, but also act as influencers with an important social function. They select which products and special promotions will be offered on their local network, based on their own experience and knowledge of the needs of their group members. In effect, they are sales personnel, who test, recommend and curate products on behalf of their neighbours. Moreover, the informal nature of the sales process – which is usually conducted through group discussions – means that it is also a place to exchange thoughts, to share information and to get to know your neighbours better. This perfectly matches the more social nature of online shopping in China.

The battle for the second-rank cities

Group purchasing opens up access to a market of millions in secondary and tertiary cities, which until now have been hard to reach. About 85% of the registered group coordinators live in this kind of city. The statistics leave no doubt that the potential is huge. The Kantar research bureau estimates that income from group purchases trebled between 2018 and 2020, reaching 13.6 billion dollars in that latter year. In September 2020 alone, 101 million people in China bought fresh products via WeChat mini-programmes, the majority of which are apps for group purchases. This was an increase of almost 70% in comparison with the previous year.

China's largest tech-companies are fighting hard to get a share of this market. JD.com announced that it has invested no less than 700 million dollars in the Xingsheng Youxuan group purchasing platform, following in the footsteps of Tencent, which had previously pumped 100 million dollars into the same company. Market leader Alibaba has invested almost 200 million dollars in a different platform called Nice Tuan.

That being said, smaller players are also hoping to get a slice of the cake. During the corona crisis, Pinduoduo (another popular app for group purchases) launched Duo Duo Maicai, a delivery service that made it possible for customers to collect their ordered goods from neighbourhood stores the next day. This initiative helped the app to break through the barrier of 750 million users and allowed the company to make a profit for the first time in its history. Didi, the Chinese answer to Uber, likewise threw itself into the battle and immediately attracted 550,000 orders per day in three cities.

A threat to physical retail?

Current estimates suggest that almost all the main players are making a loss. Although for the time being they try to explain that this is a 'conscious choice', following the example of Amazon and other disruptors in the early e-commerce years, it remains the case that small margins and heavy investment make this a market segment where it is very difficult to make money, particularly bearing in mind that prices are being kept deliberately low in an effort to break open the market. Even so, Didi top man Cheng Wei readily admits that no cost is too high and no pit too deep to become the number one in this sector (Lee & Jiayi, 2021).

Several wholesalers and independent traders have already been driven into bankruptcy, which has prompted the Chinese authorities to investigate possible dishonest practices and abuses. The government has drawn up a list of restrictions to apply to companies that are active in group purchasing, such as a prohibition on dumping prices to eliminate the competition and a ban on false advertising with reduced prices and misleading product information. These measures reflect a fear that group purchasing platforms could put many local grocery stores and perhaps even physical supermarkets out of business.

Easier than a convenience store

Faced with all this online competition, does the physical store still have a function? Walmart believes that it does – but this function will have to be radically different from the past. Because online grocery shopping is so successful, the CEO of Walmart's Chinese operations thinks that even local convenience stores are destined to disappear in the not-too-distant future. 'When you can have online deliveries made to your home within 30 minutes, the convenience store is no longer all that convenient. It will cost people more time to shop in a physical store, no matter how close it is.' (Retailhunt China, 2019). Even so, Walmart want to continue stimulating online customers to visit its physical supermarkets, where they will be able to find greater personalisation and a top-class experience. Physical shopping must become fun shopping for food too. In this way, every channel can play a different role for the consumer.

It is an interesting argument, but not one that prevents other players from continuing to invest heavily in physical convenience stores, although there is a general acceptance that these must become ultra-convenient. In this respect, China is even ahead of Amazon when it comes to the development of high-tech shops that take up an absolute minimum of their customers' time. JD.com's 7Fresh chain uses smart shopping trolleys that 'guide' customers through their shopping list. At Freshippo, self-scanning and payment with facial recognition are already standard.

Unmanned container stores

While the first unmanned stores are only just starting to appear in Europe, in China they have been part of the shopping scene for more than five years. The pioneer was BingoBox, founded in 2016, which makes use of various technological tools to offer consumers a seamless and contact-free experience. One of the most ingenious of these tools is an image-recognition box that looks like a microwave. You simply put the products you want into the box and it recognises them, adds up the bill and lets you pay (contact-free, of course)! You are out of the store in minutes.

Meanwhile, Auchan is also experimenting with unmanned and checkout-free containers, as is JD.com with its X-stores, which are open 24/7 and work with RFID labels on every product. In this way, the purchased products are scanned at the exit doors, with the cost being automatically deducted from the customer's bank account via facial recognition.

Yet for all these clever initiatives, a real breakthrough is still some way off. Most unmanned stores make a loss and many have already closed. Glitches in the systems mean that the promised high-tech 'ease' fails to deliver fully, while competition from the numerous physical local stores remains fierce.

Food as an element of power

Paradoxically, it is these same independent local grocery stores – the so-called 'mom & pop' stores – that are profiting from this evolution. Unmanned shops might not be a great success, but the lessons learned from them are not being lost. The two dominant players in the Chinese market, Alibaba and Tencent, now understand that they too can be digitalised, allowing the tech-giants to at last get a grip on the smaller cities and even the countryside. These stores can serve as local collection points, service points and fulfilment centres. Their stock can also be digitised, so that they can even acquire their own webshop.

By affiliating these millions of small store holders, the large retail groups can effectively open hundreds of thousands of new shops and expand their ecosystem exponentially. These ecosystems are the key to success – and to power. As with Amazon, Alibaba and Tencent are not interested in food per se, but in the snowball effect that food has on total consumption patterns. As a result, food is the battleground for a fierce struggle for supremacy between the two most powerful forces in the Chinese market.

It is interesting to note that all the major retailers work with one or the other, but not both. The division of the market began in 2017, when Alibaba (then the local number two) took over RT-MART. Since then, Auchan has also sold all its Chinese activities to its former partner, Alibaba, while Carrefour concluded a collaboration agreement with Tencent, although, like Auchan, it has recently thrown in the towel. That is how the current culture in China works: no-one is big enough or strong enough to go it alone, so it is necessary to make strategic partnerships. This is one of the reasons why innovation and growth happen so fast. At the same time, it also explains why the Chinese government keeps a close eye on the resulting duopoly and is taking more and more measures to curb the huge power of the two technology giants.

Expansion to Europe

In recent times, growing resistance (and maturity) in the domestic market has prompted the Chinese leviathans to explore new horizons, in the hope of making fresh conquests. In terms of food retail, it is JD.com that has so far taken the most concrete steps with its 7Fresh concept. An examination of vacancies reveals that JD.com has plans to open supermarkets in the Dutch capital Amsterdam in the near future. Fresh food, the development of a store network and a combination with e-commerce seem to be the priorities. A huge warehouse facility will also be opened in the southern part of the country at Venlo, a strategic location close to the borders with neighbouring Belgium and Germany. France is also within easy reach.

In the meantime, Alibaba has also launched its European offensive with a distribution centre in Belgium and physical AliExpress stores in Spain. To allow its international business to grow further, the company is also developing a huge network of influencers and content creators, whose task is to introduce and stimulate livestreaming and shopping via social media in the European market. For the time being, Alibaba has made no move into the European food market, preferring to concentrate on the domestic Chinese market, where its aim is to increase access to European grocery products, which are still seen as a guarantee of quality in China.

Technology producer AiFi, developer of smart cameras for autonomous supermarkets, also already has a foot firmly in the door of the European market. It was instrumental in creating Ahold Delhaize's checkout-free convenience concept AH to Go and the automated stores of Wundermart, which aims to open more than a thousand of these unmanned outlets.

It is worth noting, however, that every Chinese company operating outside of China always brings with it a piece of the Chinese state. Unhindered by GDPR legislation, it is not just the Chinese tech-giants but also the Chinese government who are using their technology to discover to what is going on in Europe. Acquiring data is the key. Of course, this is equally true of Amazon and the other GAFA (Google, Amazon, Facebook and Apple) companies, although they do so without state interference.

To expand Amazon in Europe, Jeff Bezos usually starts via partnerships, until Amazon has learnt to stand on its own two feet, when it then turns on its former partner and becomes a direct competitor. This is what happened in the UK, for example, where Amazon entered into a partnership with the Morrisons supermarket chain. Morrisons sells part of its assortment via Amazon, but this has not stopped the Seattle-based shark from starting to open its own supermarkets in England. Experts think it is only a matter of time before Amazon does a UK takeover similar to its Whole Foods acquisition in the US. This is probably not good news for Día and U2 Supermercato, Amazon's local retail partners in Italy and Spain, where the Bezos empire is also planning to extend its fresh food activities.

When two dogs fight over a bone ...

Will European retail soon be overrun by the Americans or the Chinese? Or will it be a case of when two dogs fight over a bone, a third dog runs away with it? The Russian chain VkusVill has begun its own campaign to conquer Europe by opening stores in Amsterdam and they also have France in their sights. Their ambition: 'To permanently change the West European supermarket landscape with healthy food and convenience,' says founder Andrey Krivenko (Neerman, 2020). He argues that classic retail can no longer meet the needs of modern consumers. In its own domestic market VkusVill is achieving an annual growth in turnover of an impressive 60%, coupled with the opening of an equally impressive two new stores per day, and all without external finance.

But VkusVill is not alone. The Russian discounter Mere, a real price-breaker, also has major ambitions in Western Europe. After Romania, Germany, Poland, Spain, Lithuania and Latvia, Belgium, France and the UK are now on its list of targets. This aggressive retailer sells a limited assortment of 1,500 to 2,000 essential products, displayed on pallets and at prices that are 20% lower than you can find at the traditional European hard discounters.

Of course, it doesn't stop there. Other challengers are also quietly making their preparations. They prefer to keep out of the limelight, but behind the scenes they are busily pulling various strings that give them access to all-important data. Instacart and Ocado are both companies that help physical distributers with their online fulfilment. Instacart does this literally, by collecting orders for consumers from supermarket shelves. Ocado does it by taking the full e-commerce process out of the hands of the retailer. Everything, from the software platforms to the physical fulfilment infrastructure, is under their control. They have the software, they have the logistics and they know what – and, just as importantly, who, where, when and how – consumers order. In other words, they have real power. Little wonder, then, that a desperate Marks & Spencer was prepared to pay 750 million pounds for just a 50% participation in Ocado.

WHAT CAN WE LEARN FROM CHINA?

Whether you call it omnichannel, O2O, OMO or new retail, it all boils down to the same thing: the seamless integration of the online and offline channel. In terms of the digitalisation of the food sector, this means combining the best that physical stores have to offer with the convenience of e-commerce. In the long term, this is not only the best mix for the consumer, but also for the retailers. Walmart has proven as much in China: in just nine months its dark stores at strategic locations were breaking even. Amazon Fresh also quickly learned the necessary lessons from its first real O2O concept.

Urbanisation strengthens e-commerce

Population density is a crucial factor for the success of these kinds of model. The high property prices in large urban centres play into their hands. The massive influx of people into the cities causes these prices to rise, so that the price per square metre becomes too expensive to set up physical supermarkets in these metropolitan locations. At the same time, the problem of mobility in cities is increasing. If people cannot get to the stores, the stores have to get to the people. Amazingly fast delivery by bicycle or motor bike is now possible in most major world cities. Automation and robotisation will ensure that such things are soon also possible in smaller cities.

However, the success code for a fully digitalised chain has not yet been cracked, not even in China. But it is not for want of trying. Typically Chinese short-chain initiatives such as group purchases via social media (WeChat), livestreaming and digital farmers' markets are evolving quickly from marginal phenomena to new disruptive forces, to such an extent that the Chinese government has become suspicious of group purchasing. It is only a matter of time before the West seeks to find some equivalent initiatives of its own.

Diversification of store formats

In the middle of all these developments, the role of the physical store is also changing. For example, neighbourhood stores are an ideal option for strategic but over-expensive locations, preferably checkout-free and unmanned to ensure optimal time and cost-efficiency. But in future every store format will need to match a specific profile. So, too, will warehouses. If they wish to operate efficiently in the cities, retailers will need to work with dark stores and dark kitchens alongside their existing store network.

These facilities can be housed in non-commercial spaces, which are likely to be much cheaper. For example, old car parks and disused office premises are ideal. In particular, the potential for grocery goods and non-cooled products is huge. Moreover, once they possess the right data, companies can create the right assortment with a high degree of certainty. As a result, the products will be located where local demand is greatest.

Whether we are talking about hybrid store/warehouses or full cloud stores, the alliance between data and technology is crucial. Compiling orders at Walmart stores in China takes an average of ten minutes, but four-minute completion is nothing out of the ordinary. Technology and data optimise both the picking order in the store and the delivery route. As a result, retailers can succeed in getting orders delivered, even in congested urban settings, within an hour (in fact, the average time is forty minutes). The message is clear: everything the retailer can automate must be automated, if they want to remain competitive in the future.

Open mini-platforms

Convenience stores in combination with e-commerce will become the twin key drivers for the food retail of tomorrow. For smaller purchases, consumers now expect ultra-convenience from highly automated local stores. Customers who are known simply walk in, take what they want, and walk out again, with payment being made automatically. For the larger weekly shop, they order online with home delivery. Once again, data is the key for moving towards more personalised service provision. Sales data will make it possible to fully automate repeat purchases, so that the customer doesn't even need to go to the store. This has been a dream of Amazon for years, but subscription models are now also becoming the focus of interest for other retailers.

In this way, supermarkets can become open mini-platforms, to which other players can also hitch their wagon. In China, retailers have already understood that they need to be part of an ecosystem – in their case, one of the systems controlled by Tencent or Alibaba – if they want to have a chance of securing a share of the market and a share of stomach. E-commerce specialist JD.com, a branch of the Walmart family, bought its way into the Tencent ecosystem, as a result of which the company now benefits from the system's technological and logistical support, while its hypermarkets now also offer physical local hubs for non-food articles. The motto 'if you can't beat them, join them' necessitates the conclusion of strategic partnerships, even with competitors, if needs be. This is sometimes referred to as coopetition! It is better to be a small follower in a winning platform, than the leader of the losing side.

The future of shopping according to JD.com?
The retailer explains its vision in this remarkable video:

https://youtu.be/sJmIqA3TPrg

ONLINE

Trendsetters for the future

How food futurist Mattia Crespi envisions the future of our food ecosystem

When futurist, technology strategist and innovation expert Mattia Crespi starts to philosophise about the future of our food, it is sometimes difficult not to feel just a little bit giddy. His work concentrates on emerging digital technologies, virtual and augmented reality, the evolution of the internet and the future of our world ecosystem. On behalf of the Institute for the Future in Palo Alto, he compiled his study entitled *Food Innovation: Recipes for the next decade* (2017), in which he outlined five trendsetting 'recipes' that will give the future of food added spice and flavour.

1. Scalable diversity

We are understanding more and more about the way microbes work. As a result, scientific innovation can make it possible in the years ahead for microbes to become an integral part of what and how we eat. According to Crespi, microbes will no longer be our enemies; on the contrary, they will be our allies. An in-depth knowledge of these microscopically small organisms can help us to improve food safety and add additional flavour to our food. New insights into the functioning of microbes have also revealed that each person reacts differently to specific foodstuffs. In time, this understanding will make it possible to develop highly personalised diets for each individual. At the same time, microbes can help to create greater biodiversity. Recent research has shown conclusively that a varied plant-based diet, in which microbes play a part, is essential for a healthy lifestyle.

2. Experimental biodesign

Thanks to scientific advances in the field of biology, we are now able to adjust the composition and properties of various foodstuffs. From cultured meat to molecular cooking, biology and design complement each other in a shared passion to create innovative food. New possibilities – often at the intersection of gastronomy, science and technology – will accelerate the development of new flavours, new culinary experiences and even new food systems. Whether we are talking about sustainable food or new multi-dimensional taste and texture sensations, biodesign (the integration of design and biology) and synthetic biology (re-writing the genetic code of organisms) will ensure that the future sees the discovery of fundamentally new types of food.

3. Rewritable stories

From today's closed stories about what we eat, spread by government authorities and the food industry, Crespi argues that the narrative about food needs to move towards more open stories that are constantly in movement, to which each community and each individual makes a contribution. The importance of these stories will increase, because sometimes it is the only way to differentiate yourself from others, if you are a food provider. The experience and the story will become the core value of foodstuffs, but no-one will have a monopoly on those stories. They will be constantly tested against reality by consumers – now armed with sensors, the internet and the experiences of their network – and, where necessary, will be rewritten by them and shared. This is in keeping with the new bottom-up culture and the necessary transparency in a world of empowered consumers. Contributing to these open stories and stimulating their spread is a good way to build more authentic relationships between people and brands.

4. Intelligence in the cloud

We are evolving towards decentralised and efficient systems for the management of our food systems. Agriculture, but also the rest of the food value chain, will become automated to a significant degree, thanks to the use of intelligent technology. Robotics and connected devices will constantly learn, so that they can measure and act with increasing precision. What used to be largely a matter of intuition, from the adjustment of flavours in recipes to the monitoring of crop growth, will become a matter of science: measurable and controllable.

Technological automation will also play an important role in the marketing of food. A new economy will emerge in which trading transactions, from sales negotiations to the adjustment of the offer, will happen without human intervention. The use of data – ranging from changing consumer expectations to fluctuations in the climate – will make the food market more responsive than ever. Automation will even play a role in the relationship with the consumer: networks of household utensils and portable devices will communicate directly with food producers and distributors. Knowledge sharing and data collection in cloud systems will ensure that food, people, tools and data find each other and continually strengthen each other in worldwide networks. According to the Institute for the Future (IFTF), this new technology will be low threshold and relatively cheap, so that it will be available even to smaller players.

5. Engaged eaters

The food system will be reinvented by consumers, possibly in collaboration with the sector (although possibly not). Consumers sense that they have lost their connection with and confidence in the food industry. They want to restore their lost contact with the things they put in their stomachs and will do this either by once again becoming producers themselves or by seeking to collaborate with existing producers. By 2025, consumers will be much more than simply the demand side at the end of the food chain, says the IFTF. They will help to lead the evolution towards a new food ecosystem that embraces the values of sustainability, health, social orientation and pleasure. In the following decades, newly empowered voices – again with the help of smart technology – will make better informed consumption choices and will quite literally assist in the making of new food in a co-creation process with the producers. Distant factories will be replaced by local manufacturers, who will want to eat in keeping with their identity and will want to play an active role in making this possible. The traditional capital structures of the food industry will also change: producers will increasingly look for and find bottom-up financing from users through crowdfunding and kick-start initiatives. Consumers will also have a greater influence on content, now that the food companies see them as a source of support and even as co-workers, who they want to involve in making decisions about the food they make and sell.

Towards a future-proof food policy

The food sector is in the middle of its fourth industrial revolution. Following the phase of digitalisation and connectivity, we are now entering a world in which systems will operate and self-optimise autonomously and automatically, based on data, cognitive analysis, artificial intelligence and the industrial internet of things. A world where agility, flexibility and real innovation will be central. When products and the means of production communicate in a network, new opportunities can be created for value creation and real-time optimisation. Systems in which the cyber world and the physical world merge can generate the capabilities that are necessary for smart agricultural enterprises, factories and retail concepts.

Food is no longer a product. It has become a service. However, food as a service is about much more than simply having a meal delivered to your home. As a result of technological innovation, digitalisation and automation will revolutionise traditional business models. Everyone will be involved, because the new food ecosystem – like all ecosystems – will be a place where everyone and everything is connected. When that connectivity is disrupted, the system fails. We can see that today with regard to different links in the food chain: surpluses and shortages in agricultural supply, farmers weighed down by the burden of their debts, retailers and manufacturers who are furiously engaged in a race to the bottom, an industry that makes its consumers sick with tempting but wholly innutritious food ... Last but definitely not least, the pressure on our planet, which can no longer meet the demands of a growing and hungry population, is unsustainable.

Notwithstanding the increasing interest in short chains, with local production and consumption, it is impossible to ignore the fact of globalisation. This has resulted in a massive enrichment in the range of available food, but has made the supply chain in the food sector extremely complex and vulnerable. The entire system threatens to collapse when something unexpected happens, such as a natural disaster, a pandemic or even a wrong manoeuvre in the Suez Canal. Changing patterns of consumption in developing countries lead to price fluctuations and market disruptions. When this occurs, the consequences are felt around the globe. Hungry China is insufficiently self-sufficient in food and scours the world for scarce resources and agricultural products. The demand for pork continues to rise, as does the demand for soya, grain and maize. The traditional tea-drinkers have discovered coffee. Chinese investors are even starting to colonise French wine estates. The balance of power is shifting: Europe is at risk of being sandwiched between China and the US, while the South is now also demanding its place in the sun. Food is now politics and, as such, it needs policy.

The food chain moves into the cloud

How can we repair the connections that have been broken? In this book we explored different possibilities. It all starts with the farmer. Agriculture is undergoing a technological and digital revolution: drones, robots and self-driving tractors are taking over much of the work; sensors and smart cameras now monitor both crops and livestock; the analysis of big data predicts yields and makes it possible to respond more flexibly to market demand. New production methods are increasing productivity and are finally making the agricultural sector climate positive. This revolution now needs to be transferred in an appropriate manner to the South, where farmers often still work in primitive conditions to harvest the raw materials and foodstuffs that we regard as indispensable, but for which we in the North are not willing to pay a fair price.

The food factories of the future will be highly digitalised production units, capable of providing healthier and safer products in a more efficient manner, thanks to the use of automation, AI, the internet of things and cloud computing. The factories will also work more flexibly, allowing them to viably produce nutritionally modified or even personalised food in smaller quantities. Crucially, the digital transformation will finally make it possible to achieve real transparency throughout the entire food chain, from farm to fork and even beyond, because what happens after consumption – the processing of surpluses and waste – is also an integral part of the product cycle.

Another factor of vital importance is consumer confidence. As a result of various food health incidents and scandals at the start of the century, people now distrust agriculture and the food industry. Repairing this breach of trust is the task of every link in the food chain. Consumers now expect openness about ingredients, methods of production, working conditions, climate impact, health benefits, profit margins, etc. With the advent of sophisticated and high-tech food solutions, such as fermentation, cultured meat and personalised food, these expectations will become ever stronger. Food goes into the mouths of consumers: they need to once again have faith in what they are eating.

E-commerce in food will continue its breakthrough, certainly for run shopping and repeat purchases, for which the traditional shopping list will be replaced by subscription systems and automation (think, for example, of the smart fridge that orders its own refilling). Manufacturers and chefs will have a whole range of possibilities to promote and deliver their products and meals to hungry consumers, at every opportunity and at every moment of the day. Moreover, this range will be increasingly adjusted to meet the specific needs and wishes of individual consumers: lifestyle, life phase, health, medical condition, etc. will all be taken into account.

The food retailer will not disappear, but will also inevitably undergo a far-reaching transformation. The dominance of the supermarkets is destined to come to an end before too much longer. Physical stores will take on many different forms and fulfil a variety of different functions, from quick last-minute stores to last-mile solutions and even catering. They will also become more closely knit into their hyper-local market, serving as a mirror for the various segments of the population that they serve, which immediately adds a new dimension to category thinking. Assortments, shelf planning, promotions and communication will all be radically different, depending on the store location. 'One size fits all' will be a thing of the past. Large supermarkets and hypermarkets will also need to make better use of their excess space and, consequently, they will need to drastically revise their business model. Automation and digitalisation will set the tone.

Food stores must learn to relinquish old parameters like turnover per square metre or productivity per man hour. In future, it will be omnichannel efficiency that counts, linked to experience per square metre as a new parameter: if retailers fail to provide an experience, people will simply shop online. The digital revolution offers food retail the opportunity to reinstate and revalue the 'market' element of the supermarket concept. The stores of the future must be connective platforms, where shoppers can find inspiration, information, events, cooking classes, product demonstrations, etc., but also where they can search for, find and order online the things that are not available offline. But stores must also become places where brands and suppliers are given access to a specific public, to shop space and logistical services, to data and market intelligence, to co-creation.

As a result, the trading relationship and balance of power between producers and distributors will change significantly. Unnecessary intermediaries will disappear. Market demand will steer production much more directly. In a blockchain-managed platform economy the food chain will move in part into the cloud. Data transparency must become the norm.

Gigantic challenges

Will technology solve our problems? Far from it. Technology is an enabler, not a solution. Solutions require new organisational structures and new company cultures. Succeeding in the new environment of the future means creating new partnerships, finding the right speed, developing talent and setting priorities in a world of increasing complexity. This will only be possible with open communication, hunger for innovation, continual professional learning and development, data-based decision-making, and trust in processes and IT systems. Trendsetting companies will invest in cloud-based platforms that connect the databases and systems of different teams, so that data silos disappear and the personnel of different departments have access to the same shared information, which they can interpret and process in real time. This transformation will result in a demand for new skills, so that we can expect a hard-fought war for talent.

There will also be a need to come to terms with some harsh economic realities. Everyone in the chain still wants to earn money and everyone in the chain wants to cut costs. Competition amongst the retailers remains fierce. They have no option but to negotiate hard with their trading partners, so that they can offer their price-conscious customers attractive deals. Nor should we forget that there are still people, even in the rich West, who go to bed hungry each night. Of necessity, they need to keep a close watch on their expenses. They are not interested in expensive organic food, plant-based burgers or fair trade products, never mind 3D food printing and apps that give you recipes based on your DNA. Moreover, this vulnerable part of the population seems likely to increase rather than decrease, and this at a time when the risk of a post-corona economic crisis is real: the pandemic had a massive and destructive impact on many companies and sectors.

Society also has its role to play. We need to make important choices. What kind of food system do we want tomorrow? We realised some time ago that food is too important to be left to the forces of the free market, but there is no general agreement about the concrete actions that need to be taken as a consequence of that realisation. The reformation of European agriculture is taking shape, but is causing friction and still needs to be embedded in a global economy where conflicting interests make it very difficult to organise trading relations between the different food chain actors in a manner that is transparent and fair. Change always costs money. So how are we going to split the bill? The European Green Deal foresees budget for research and innovation, and supports farmers who want to make the switch to sustainable agriculture, but is that enough? Technological innovation and digitalisation can certainly improve the efficiency of the food chain, but it is still not clear how the costs and benefits can be equitably shared.

It is also significant that liberal governments are now announcing measures to push both the food industry and consumers in the direction of healthier eating patterns. For a long time, this was a taboo subject, but in future the production, promotion and sale of food products that are regarded as unhealthy will be subject to the same kind of restrictions that currently exist for the marketing and consumption of tobacco and alcohol. Education and awareness campaigns will not be sufficient: regulation is necessary, if we seriously want to tackle the urgent and increasing problems of obesity, 'civilisation illnesses', allergies and intolerances.

The challenges in the food system are structural. It will require a huge effort to deal with them successfully. Everyone will need to leave their comfort zone. Uncertainty about the future has never been greater than it is today. At the same time, the possibilities to make different choices have also never been greater. The cards are favourable for companies that are prepared to fundamentally question what they do and are willing to reinvent themselves. We have an opportunity to take a giant step forwards towards a fairer, healthier and more sustainable food system. We cannot afford to miss it.

A historic turning point

In the years to come, we must never forget that food is about so much more than the straightforward delivery of the calories and nutrients we need to survive. If that was the case, pills and a tube of astronaut food would suffice. But we humans could never be satisfied with that. Food connects us. Food brings people together. Around a table, sharing a meal or a drink, we cement relationships, share experiences, tell each other stories. And we enjoy new culinary sensations and surprising ingredients, which provide us with experiences that make us happy and fulfilled.

Similarly, running a physical store or launching a food brand on the market are not purely logistical operations. People derive meaning from what they buy and where they buy it, even when they are only shopping for their daily groceries. The store experience at Aldi is not the same as the store experience at Whole Foods. A pack of private label coffee is not the same as drinking at Starbucks. Just as every retailer and manufacturer will be required to become a technology company, a data company and a media company, so they will also need to become experience companies. In other words, a brand. And the higher their brand value, the higher their market value. It is simply a question of survival. If you cannot offer added value, you will be dead in the water in a digital world where the major platforms copy everything.

This presupposes 'purpose'. We expect brands to take their social responsibilities seriously and to an increasing extent. They need to take clear standpoints that do not hide behind corporate jargon. Consumers vote with their wallets and purses every day, and each purchase transaction is about more than just money: it is about data, lifestyle, choices and values. This applies equally for every contract that a purchasing director concludes with a supplier, for every campaign that a marketeer launches, for every innovation that a manufacturer develops, and for every display that a retail manager sets in his store.

Brands no longer sell products, but ideas. They no longer satisfy hunger, but desires. Brands have the chance to become the motors of change, the driving force behind a new, sustainable, honest, transparent, healthy and enjoyable food system. The momentum is already there. What is happening today has the potential to change everything: the way we do business, the structures of our society, the way our politicians govern, the way we work, consume, spend money, travel and educate. We are standing at a turning point in our history. Let us seize this golden opportunity with both hands.

Takeaway: five ingredients for a new food system

1. An agricultural revolution

Everything begins in the fields. The agricultural sector is facing a double challenge. On the one hand it needs to become more productive, while on the other hand it needs to become more ecological. The agriculture of tomorrow must no longer exhaust the soil or use harmful chemicals. Instead, it must restore biodiversity and become circular. Farmers must manage to grow more food more sustainably on the same area. Technology will take over part of the hard manual labour. Artificial intelligence will optimise cultivation. Urban agriculture and indoor vertical farming will bring production closer to the market. However, it needs to be remembered that the challenges in the poor (and warm) South are not the same as in the rich and densely populated West.

2. The transition to plant-based

Western society needs to be cured of its addiction to meat. This does not mean that veganism must become the new norm. But the daily eating of meat – often more than once a day – is an aberration that is unsustainable in the long term. This trend is already on the move: more and more consumers are describing themselves as flexitarians and are adding more plant-based variation to their diets. This is becoming easier, because the food industry is increasingly developing tasty and healthy alternatives based on new ingredients and technologies. A breakthrough in the production of 'clean' meat in bioreactors is getting closer, but still has a number of obstacles to overcome.

3. A data revolution

Transparency is an absolute prerequisite for a truly fair and efficient food system. Data streams can highlight problem links in the supply chain and help to keep supply and demand in balance. Data will feed and direct the factories of the future. Data will objectify the difficult trading relationship between retailers and suppliers. Data will help us to eliminate food waste and loss. Data will personalise our food offer to reflect our hyper-individual health needs.

4. A retail reset

The shortest route between farm and fork will no longer necessarily be via the classic supermarket. New distribution models will appear to satisfy the different needs of an increasingly fragmented and hyper-diverse society. In the omnichannel retail world of tomorrow, hybrid stores will fulfil different functions, while online and offline will be seamlessly integrated. Far-reaching digitalisation will lead to global consolidation, in which a limited number of platform giants will acquire a disproportionate amount of market power. However, new partnerships and innovative forms of coopetition will provide opportunities for fighting back.

5. A global paradigm shift

All these necessary transformations presuppose a new story, a unifying project for the future. The corona pandemic might actually have helped to speed up this process. As a result of the crisis, there is a growing awareness amongst consumers about the impact that their purchasing decisions can have on society and the environment. Brands and companies are under increasing pressure to reduce their ecological footprint, to achieve the UN's Sustainable Development Goals, and to accept social responsibility for their actions. Governments are taking measures, both locally and at the European level, which is by no means self-evident in the context of a global market in which the balance of power is constantly shifting. Even so, collaboration between all the different actors at every different level will need to be governed by new models. In a fluid society there is no longer any place for classic top-down decision-making, rigid hierarchical organisations and strict orders. Organisations will need to reinvent themselves, developing new relationships and fresh connections. This will also necessitate different skills from their personnel. The future belongs to open networks, decentralisation and transparency. It is only with these ingredients that we can together prepare a new recipe for the world of tomorrow.

'The financial question is crucial for a successful transformation'

We showed our book in advance to Frans Muller, CEO of Ahold Delhaize, and asked him for a reaction. He emphasised that it is important to understand the complexity of the situation we are facing.

'The challenges in the food transformation chain are major and affect many different fields: climate, food security, ecosystems, waste, social relations, etc. For this reason, it is important to take a broad view of the entire chain, in all its aspects, from farmer to consumer. That is also an essential condition for a better balancing of supply and demand.'

'In addition, it is important not to lose sight of economic realities. Transformation costs money. So who is going to pay? Everyone in the chain wants to make a profit: the farmer, the manufacturer, the retailer. There are competitive factors at play and we also need to remember the purchasing power of the consumer. For this reason, transformation can only be successful once some of these elements, such as the financing of climate improvement measures and the stimuli for regenerative agriculture, are regarded as precompetitive.'

'Financing and support will be crucial to the transition towards a more sustainable food system from farmer to consumer. This is where government can play a role by creating a level playing field. The European Green Deal with its 'Farm to Fork' strategy is a good start that can help to get a movement for change under way, although it is also possible to criticise some of the choices made.'

'Another equally important task is to make consumers more aware and to restore their confidence in the food sector. The sector will need a convincing plan of action, open and honest communication, and the ability to demonstrate that it can be a reliable partner. We must be able to show that we are there to guide consumption choices, not dictate them. In this respect, I see a key role for the standardisation of information, with nutritional labelling (the Nutri-Score, Guiding Stars, etc.) and, later on, even climate labelling.'

'Climate themes and environmental dilemmas remain an urgent concern. We must strive to achieve a (net) zero growth in greenhouse gas emissions, the recovery of biodiversity, and the regeneration of healthier soil. Technology and digitalisation can help us in these matters. Upscaling and technological development will go hand in hand with authenticity and naturalness. Finding substitutes for animal proteins is an important factor in helping to repair the damage done to the climate. Yet even with all these necessary changes, I believe that the omnichannel supermarket will continue to be the most important sales channel for the largest part of the consumer market.'

'The COVID crisis has strengthened awareness of a number of issues. People are now cooking more at home and have more respect for food and its production. Convenience is also becoming more important and health now occupies a higher position on the social agenda, as does climate change. '

A final question: what does Frans Muller hope to see on his plate in 2030? 'Providing we can ensure a sustainable stock of fish by that date, I would like to see a nice sole, turbot or brill!'

// Jef Colruyt / CEO Colruyt Group //

'As retailers, we can continue to play an important role in the future of food. We must continue to innovate, constantly reinventing ourselves in the interests of the health of the consumer, whilst at the same time taking account of factors like the environment, animal welfare and society as a whole. Making a positive difference step by step is (and will remain) the message. This book shows that the future will be challenging, but also full of opportunities that we hope to seize with both hands, together with our different partners.'

// Nils van Dam / CEO Milcobel //

'In *The Future of Food* we are taken on a fascinating and inspirational journey into the world of food. For those who believe that the food market will change very little or even not at all, it will be a real eye-opener. Jorg and Stefan succeed in explaining clearly the complex food ecosystem, where everything is connected to everything else, and they provide us with surprising insights into possible future scenarios. I would recommend this book to anyone and it is a must for every food professional. Enjoy your read!'

// Alexandre Bompard / Chairman & CEO, Carrefour Group //

'Jorg Snoeck and Stefan Van Rompaey explore all the angles of the complex equation that the players in the food chain are now confronted with: how can we combine higher quality, respect for the environment and attractive prices all while investing massively in new distribution channels – particularly digital. This is what makes our job as retailers so difficult and noble, and why Carrefour made the 'food transition for all' its raison d'être. On its journey the challenges are as immense as they are exciting, and this book serves as a superb introduction.'

// Koen Slippens / CEO Sligro Food Group //

'*The Future of Food* is an interesting summary of all the developments and challenges associated with the food chain and what they might mean for food retail and food service. Well worth a read.'

// Hein Deprez / co-CEO Greenyard //

'For me, this is the most complete book I have read so far about new food trends and the evolution of the sustainability aspect in the production of new food products. In addition, the book also places these trends in the context of the rapidly changing manner in which food finds its way from the farmer's field to the consumer's plate. All the new technologies and ideas that are today being studied and implemented worldwide are clearly described with appropriate links to concrete examples. The knowledge that I have acquired as a result simply serves to strengthen my conviction that retailers whose footprint has a strong local anchor (and a well-spread and dense network of stores) will be capable of putting new and sustainable breakfasts, lunches and dinners on people's tables in an efficient and sustainable manner. The retailers who can succeed in incorporating new products and technologies into their business models will win the attention of consumers and become champions in completing the last mile. I am making this book compulsory reading for the top 250 people in our company.'

// Dirk Van den Berghe / board member and advisor, former Executive Vice President and CEO Asia, Canada and Global Sourcing, Walmart Inc. //

'It is clear that almost every sector is currently undergoing major technological change and experiencing the effect of far-reaching social developments. This applies equally to the food sector, but our sector is attracting more attention than most, not only because of the major impact it has on society but also because of its importance to us as individuals. Having a better understanding of what is happening now and what will happen in the future is therefore both useful and topical. The authors give their readers a fascinating insight into the food world of tomorrow, which can be very positive for the food sector, providing all the participants in the value chain play their role to the full. This book is highly recommended for anyone who is passionate about the future of food.'

// **Wim Destoop** / VP General Manager PepsiCo North-West Europe //

'*The Future of Food* has it all. The book deals with all the challenges and opportunities facing our wonderful sector. It reads like a thriller for everyone who is passionate about food. It covers a wide range of themes that can give pause for thought to every manufacturer and distributor, whether great or small, local or international. The excellent examples illustrate the complex revolution and transformation that is taking place in the world of food. Food is emotion. Consequently, it is more important than ever to understand consumers, so that together with all other relevant parties we can find the right solutions.'

// **Lieven Vanlommel** / CEO Foodmaker //

'We have known for a long time that Jorg and Stefan are top experts in the world of retail. With them, I have been involved in many fascinating retail hunts all around the globe. Everything that we discovered during those adventures has now found its way into this book in a clear and concise form. After the recent pandemic, we all have a responsibility to motivate people to live happier and healthier lives. This will require a transition from mass consumption to a more aware and sustainable form of consumption, which effectively means a switch to a more plant-based diet. This is something that we at Foodmaker can only applaud. From farm to fork, plant-based and supported by technology is the right way to avoid new pandemics in the future. Thank you for this super-interesting book and thank you for our fascinating time together!'

// **Olaf Koch** / Partner Zintinus and former CEO of Metro //

'We are at the doorstep of one of the biggest changes in the food industry. Greatly altering consumer requirements paired with the need to use our planet's resources more responsibly will lead to an unprecedented wave of innovation. Jorg Snoeck and Stefan Van Rompaey illustrate that with great clarity and precision. This book is a wonderful compendium of the various trends that will lead to exponential change. It is a powerful wake-up call but also a guide to encouraging perspectives. We at Zintinus share many of their views and are fully committed to do our best to actively contribute to the transformation of the food industry. For the good of planet, people and investors.'

// BIBLIOGRAPHY //

Adams, S. *How Daily Table Sells Healthy Food To The Poor At Junk Food Prices*, 26 Apr. 2017 [Consulted on 09/03/2021] via https://www.forbes.com/sites/forbestreptalks/2017/04/26/how-daily-table-sells-healthy-food-to-the-poor-at-junk-food-prices/?sh=463bb1201bc3

Ahmed, A. (2008) *Marketing halal meat in the United Kingdom*. British Food Journal, 110(7):655-670.

Alexander, D. *3D Printing Will Change the Way You Eat in 2020 and Beyond*, 27 Mar. 2020 [Consulted on 08/03/2021] via https://interestingengineering.com/3D printing-will-change-the-way-you-eat-in-2020-and-beyond#:~:text=To%20the%20future,will%20be%20created%20on%2Ddemand

[Anonymous] *Foodtech Brands – Top European Foodtech Startup Brands Inventing The Future of Food*, Oct. 2020 [Consulted on 03/03/2021] via https://www.digitalfoodlab.com/reports/2020-food-brands/download

[Anonymous] *Portuguese online supermarket eliminating food waste*,22 Oct. 2020 [Consulted on 09/03/2021] via https://www.theportugalnews.com/news/portuguese-online-supermarket-eliminating-food-waste/56321

[Anonymous] *Daily 'happy hour' at supermarkets in Finland*, 19 Sep. 2019 [Consulted on 09/03/2021] via https://www.retaildetail.be/nl/news/food/elke-dag-happy-hour-de-supermarkt-finland

Aravindan, A. (Reuters) *Singapore ramps up rooftop farming plans as virus upends supply chains*, 8 Apr. 2020 [Consulted on 02/02/2021] via https://news.trust.org/item/20200408062120-9zn5k/

Baraniuk, C. *The Plan to Rear Fish on the Moon*, 22 Feb. 2021 [Consulted op 28/02/2021] via https://www.hakaimagazine.com/news/the-plan-to-rear-fish-on-the-moon/

Beckett, E. L. *Ghost Kitchens could be a $1T global market by 2030, says Euromonitor*, 10 Jul. 2020 [Consulted on 06/03/2021] via https://www.restaurantdive.com/news/ghost-kitchens-global-market-euromonitor/581374/

Berg, N. & Knights, M. (2019) *Amazon: How the World's Most Relentless Retailer will Continue to Revolutionize Commerce. London: Kogan Page.*

Berg, P. et al. *The drive toward sustainability in packaging—beyond the quick wins*, 30 Jan 2020 [Consulted on 30/03/2021] via https://www.mckinsey.com/industries/paper-forest-products-and-packaging/our-insights/the-drive-toward-sustainability-in-packaging-beyond-the-quick-wins

Bhargava, S. e.a. *The young and the restless: Generation Z in America* [Online], 20 Mar. 2020 [Consulted on 16 April 2020]; via https://www.mckinsey.com/industries/retail/our-insights/the-young-and-the-restless-generation-z-in-america

Bliss, D. *This vision of food's future could have legs – but rather more than you might like*, 1 Sep. 2020 [Consulted on 24/02/2021] via https://www.nationalgeographic.co.uk/environment-and-conservation/2020/08/this-vision-of-foods-future-could-have-legs-but-rather-more

Buller, A. *UK's biggest premium halal food producer Haloodies sees 30% rise in online sales*[Online], 18 Aug. 2020 [Consulted on 16 April 2020]; via https://www.arabianbusiness.com/retail/450897-uks-biggest-premium-halal-food-producer-haloodies-sees-30-rise-in-online-sales

Candau, M. *Sustainable rooftop-fish-farming conquers Brussels city heights*, 28 Feb. 2020 [Consulted on 09/02/2021] via https://www.euractiv.com/section/agriculture-food/news/sustainable-rooftop-fish-farming-conquers-brussels-city-heights/

Carrington, D. *The new food: meet the start-ups racing to reinvent the meal*, 30 Apr. 2018 [Consulted on 24/02/2021] via https://www.theguardian.com/environment/2018/apr/30/lab-grown-meat-how-a-bunch-of-geeks-scared-the-meat-industry

Chandran, R. *6 ways COVID-19 is changing our cities*. [Online] Thomson Reuters – World Economic Forum, 8 Jan. 2021 [Consulted on 5 March 2021]; via https://www.weforum.org/agenda/2021/01/coronavirus-covid-19-urbanization-cities-change-pandemic/

Chen, L. & Zhang, T.B. red. (2020) *The Fresh Food Business. Spurring the 'Local Community' Trend Forward*. Beijing: Deloitte China.

Chong, C. *Planting the seeds for the future of farming*, 27 Nov. 2020 [Consulted on 02/02/2021] via https://www.businesstimes.com.sg/companies-markets/emerging-enterprise-2020/planting-the-seeds-for-the-future-of-farming

Chwalik, R. & Goehle, B. (2018) *The Global Innovation 1000 study* [Online]. 19 Mar. 2019 [Consulted on 18 March 2021]; via https://www.strategyand.pwc.com/gx/en/insights/innovation1000.html

Corstjens, M. *Penetration -The New Battle for Mind Space and Shelf Space*. Trigga Consulting Ltd (Oct. 2015)

De Schamphelaere, J. *Belg plant 100 boerderijen in Europese grootsteden*, 14 May, 2019 [Consulted on 02/02/2021] via https://www.tijd.be/ondernemen/voeding-drank/belg-plant-100-boerderijen-in-europese-grootsteden/10126819.html

De Schamphelaere, J. *Ex-topman Balta stort zich op verticale boerderijen*, 2 May 2019 [Consulted on 09/02/2021] via https://www.tijd.be/ondernemen/voeding-drank/ex-topman-balta-stort-zich-op-verticale-boerderijen/10122634.html

Dent, M. *Five Companies at the Forefront of the Cultured Meat Revolution*, 17 Sep. 2020 [Consulted on 21/02/2021] via https://www.idtechex.com/fr/research-article/five-companies-at-the-forefront-of-the-cultured-meat-revolution/21677

Dent, M. *Five Things IDTechEx Learned From the Cultured Meat Symposium 2020*, 19 Nov 2020 [Consulted on 23/02/2021] via https://www.idtechex.com/fr/research-article/five-things-idtechex-learned-from-the-cultured-meat-symposium-2020/22311

Dimbledy, H. & UK Department for Environment, Food & Rural Affairs. *Developing a national food strategy: independent review 2019 – terms of reference*. [Online] 29 Jul. 2020. [Consulted on 5 March 2021]; via https://www.gov.uk/government/publications/developing-a-national-food-strategy-independent-review-2019/developing-a-national-food-strategy-independent-review-2019-terms-of-reference

Dongyu e.a., *The State of Food Security and Nutrition in the World 2020*. [Online] Jul. 2020. [Consulted on 5 March 2021]; via http://www.fao.org/3/ca9692en/online/ca9692en.html

Eliaz, S. & Jagt, R. (2020). *The global food system transformation. The time to change is now*. Deloitte [Online]. Dec. 2020 [Consulted on 5 March 2021]; via https://www2.deloitte.com/global/en/pages/consumer-business/articles/global-food-system-transformation.html

European Commission, Directorate-General for Health and Food Safety. *A 'From Farm to Fork' strategy for a fair, healthy and environmentally friendly food system*. Brussels, 20 May 2020

Evans, J. &Terazono, E. *Unilever aims for €1bn sales from plant-based products by 2027*, 18 Nov. 2020 [Consulted on 16/02/2021] via https://www.ft.com/content/0a1e5e3d-a34d-44bb-a350-75f3e8700673

Evenepoel, K. *Chinese kassaloze supermarkttechnologie steekt neus aan venster in Europa* [Online]. 2 Mar. 2021 [Consulted on 18 March 2021]; via https://www.retaildetail.be/nl/news/algemeen/chinese-kassaloze-supermarkttechnologie-steekt-neus-aan-venster-europa

FAO (2020). *SDG Digital Progress Report: Sustainable Development Goal 12*. Responsible consumption and production. [Consulted on 5 March 2021]; via http://www.fao.org/sdg-progress-report/en/#sdg-12

FAO - Food and Agriculture Organization of the United Nations: FAOSTAT, 2018 [Consulted on 02/02/2021] via http://www.fao.org/faostat/en/#data/EL

FAO, IFAD, UNICEF, WFP & WHO *The State of Food Security and Nutrition in the World 2020.* Transforming food systems for affordable healthy diets. Rome, 2020.

FMI & Label Insight (2021) *Transparency Trends. Omnichannel Grocery Shopping from the Consumer Perspective.* FMI: Arlington.

Foley, J. *Where will we find enough food for 9 billion?* National Geographic Magazine. [Consulted on 5 March 2021]; via https://www.nationalgeographic.com/foodfeatures/feeding-9-billion/

Francis, T. & Hoefel, F. *'True Gen': Generation Z and its implications for companies* [Online]. 12 Nov 2018 [Consulted on 16 April 2020]; via https://www.mckinsey.com/industries/consumer-packaged-goods/our-insights/true-gen-generation-z-and-its-implications-for-companies

Funabashi, M. *Human augmentation of ecosystems: objectives for food production and science by 2045.* npj Sci Food 2, 16 (2018). [Consulted on 5 March 2021]; via https://doi.org/10.1038/s41538-018-0026-4

Future Bridge, *3D Printing and its Application Insights in Food Industry,* 11 Mar. 2020 [Consulted on 08/03/2021] via https://www.futurebridge.com/industry/perspectives-food-nutrition/3D printing-and-its-application-insights-in-food-industr/

Galloway, S. (2020) *Post Corona. From Crisis to Opportunity.* New York: Portfolio.
Geijer, T. enGammoudy, A. *Growth of meat and dairy alternatives is stirring up the European food industry,* 22 Oct. 2020 [Consulted on 20/02/2021] via https://think.ing.com/reports/growth-of-meat-and-dairy-alternatives-is-stirring-up-the-european-food-industry

Gellynck, X. *Waarom insectenburgers (nog) niet aanslaan,* 9 Aug. 2016 [Consulted on 24/02/2021] via https://www.ugent.be/bw/nl/onderzoek/ugent-crelanleerstoel/overzicht-artikels/insectenburgers.htm

Gerckens, C. e.a. (2021) *Disruption & Uncertainty. The State of Grocery Retail 2021: Europe.* Zurich: McKinsey & Company.

Ghys, I. *Kan onze landbouw klimaatneutraal?* 13 Feb 2021 [Geraadpleegd op 21/02/2021] via https://www.standaard.be/cnt/dmf20210212_98150497

Grand View Research *Online Grocery Market Size & Share Report, 2020-2027* [Online]. Apr. 2020 [Consulted on 5 March 2021]; via https://www.grandviewresearch.com/industry-analysis/online-grocery-market

Guillard V et al. *The Next Generation of Sustainable Food Packaging to Preserve Our Environment in a Circular Economy Context,* 4 Dec. 2018 [Consulted on 30/03/2021] via https://www.frontiersin.org/articles/10.3389/fnut.2018.00121/full#B68

Guterres, A. *Launch of policy brief on COVID-19 and cities. COVID-19 in an Urban World* [Online]. 2020 [Consulted on 5 March 2021]; via https://www.un.org/en/coronavirus/covid-19-urban-world

Harrabin, R. *Food 'made from air' could compete with soya,* 8 Jan. 2020 [Consulted on 28/02/2021] via https://www.bbc.com/news/science-environment-51019798

Harvey, F., e.a. *Rise of mega farms: how the US model of intensive farming is invading the world,* The Guardian [Online]. 18 Jul. 2017 [Consulted on 5 March 2021]; via https://www.theguardian.com/environment/2017/jul/18/rise-of-mega-farms-how-the-us-model-of-intensive-farming-is-invading-the-world

Heerink, M. & Bunskoek, J. *Explosieve groei wereldsteden: deze plaatsen groeien het hardst,* RTL Nieuws [Online]. 11 Feb. 2020 [Consulted on 22 April 2021]; via https://www.rtlnieuws.nl/nieuws/buitenland/artikel/5018221/groei-bevolking-grootste-steden-landen-populatie-egypte-nigeria

Henry, C.J. *Functional Foods,* European Journal of Clinical Nutrition, vol. 64, p 657–659 (2010).

Herren, H. e.a (2019). *Transformation of our food systems. The making of a paradigm shift.* Berlin: Zukunftsstiftung Landwirtschaft.

Hidrèlèy *Skirts And Heels Are Not Just For Women, This Guy Proves That Perfectly* [Online]. Nov. 2020 [Consulted on 16 April 2020]; via https://www.boredpanda.com/confident-man-wears-heels-skirt-markbryan911

Holderbeke, J. *Wat eten we in 2030? 'Een wereld die niet duurzaam is zal niet vreedzaam zijn'*, 23 Mar. 2021 [Consulted on 25/03/2021] via https://www.vrt.be/vrtnws/nl/2021/03/21/wat-eten-we-in-2030/

Hubbard, P. & Maginn, P.J. *Coronavirus could turn cities into doughnuts: empty centres but vibrant suburbs*. The Conversation [Online], 8 Dec. 2020. [Consulted on 5 March 2021]; via https://theconversation.com/coronavirus-could-turn-cities-into-doughnuts-empty-centres-but-vibrant-suburbs-151406

Institute for the Future, *Food Innovation: Recipes for the Next Decade*. Palo Alto, 2017.

Jagt, R. *Future of Food: personalized, responsible and healthy*. Feb. 2021 [Consulted on 20/03/2021] via https://www2.deloitte.com/global/en/pages/consumer-business/articles/gx-food-personalized-healthy-nutrition.html

Junjie *Pinduoduo's New Venture into Online Grocery Shopping in Quest to Expand Agriculture Share* [Online]. 13 Nov. 2020 [Consulted on 18 March 2021]; via https://pandaily.com/pinduoduos-new-venture-into-online-grocery-shopping-in-quest-to-expand-agriculture-share/

Kamel, M.A. et al. *How to Ramp Up Online Grocery—without Breaking the Bank*. Jul. 2020 [Consulted on 04/03/2021] via https://www.bain.com/insights/how-to-ramp-up-online-grocery-without-breaking-the-bank/

Lacanne, C. & Lundgren, J. *Regenerative agriculture: merging farming and natural resource conservation profitably*. 26 Feb. 2018 [Consulted on 24/02/2021] via https://www.ncbi.nlm.nih.gov/pmc/articles/PMC5831153/

Lambrechts, T. *Hoe blockchain fair trade een nieuwe betekenis geeft*. [Consulted on 27/03/2021] via https://eostrace.be/artikelen/hoe-blockchain-fair-trade-een-nieuwe-betekenis-geeft

Lee, E. & Jiayi, S. *Friendly neighbors are the key to China's community group-buying craze* [Online]. 27 Jan. 2021 [Consulted on 18 March 2021]; via https://technode.com/2021/01/27/friendly-neighbors-are-key-to-chinas-group-buying-craze/

Macready, L. e.a. (2020). *Consumer trust in the food value chain and its impact on consumer confidence. A model for assessing consumer trust and evidence from a 5-country study in Europe*. Food Policy 92 (April) 101880.

Marcotte, D. et al. *The Store of the Future: An Exploration and Planning Guide*. Coca-Cola Retailing Research Council, 2019.

Mehmet, S. *Understanding blockchain in the food industry*. 28 Jan. 2020 [Consulted on 27/03/2021] via https://www.newfoodmagazine.com/article/104104/understanding-blockchain-in-the-food-industry/

Melton, J. (2020). *2020 Online Food Report*. Chicago: Digital Commerce 360.

Murray, L. *Your Dinner is Printed*. 16 Feb. 2021 [Consulted on 08/03/2021] via https://eandt.theiet.org/content/articles/2021/02/your-dinner-is-printed/

Neerman, P. *'China heeft een voorsprong in retail'* (Jonathan Reynolds, Institute of Retail Management). 2 Oct. 2018 [Consulted on 03/03/2021] via https://www.retaildetail.be/nl/news/algemeen/%E2%80%9Cchina-heeft-een-voorsprong-retail%E2%80%9D-jonathan-reynolds-institute-retail-management

Neerman, P. *Drie grote food-revoluties op komst tegen 2030*. Jan. 2020 [Consulted on 20/03/2021] via https://www.retaildetail.be/nl/news/food/drie-grote-food-revoluties-op-komst-tegen-2030

Neerman, P. *Freshippo: de toekomst van de supermarkt, Chinese stijl*. 10 Mar. 2020 [Consulted on 06/03/2021] via https://www.retaildetail.be/nl/news/food/freshippo-de-toekomst-van-de-supermarkt-chinese-stijl

Neerman, P. *Heeft Amazon de toekomst van supermarkten nu echt gekraakt?* [Online]. 25 Feb. 2020 [Consulted on 16 April 2021]; via https://www.retaildetail.be/nl/news/food/heeft-amazon-de-toekomst-van-supermarkten-nu-echt-gekraakt

Neerman, P. *Onze bananen worden bedreigd*, [Online]. 19 Aug. 2019 [Consulted on 5 March 2021]; via https://www.retaildetail.be/nl/news/food/onze-bananen-worden-bedreigd

Neerman, P. (2019) *Retailhunt China 2018*. Antwerp: RetailDetail.

Neerman, P. *Russian chain plans Dutch, French, Belgian invasion* [Online]. 4 Mar. 2020 [Consulted on 18 March 2021]; via https://www.retaildetail.be/nl/news/food/russische-supermarktketen-stort-zich-op-nederland-en-frankrijk

Oakland Innovation. *Unlocking Personalised Nutrition*. Jun. 2020.

OECD/FAO (2020), *OECD-FAO Agricultural Outlook 2020-2029*, Rome: OECD Publishing [Consulted on 5 March 2021]; via https://doi.org/10.1787/1112c23b-en

Packham, C. *7.7 Billion People and Counting*. BBC Horizon, 23 Jan. 2020.

Parker Brady, R. red. (2018) *Appie Tomorrow. Alle ingrediënten voor een beter leven in 2025*. Zaandam: Albert Heijn.

Patel, H. et al. *Active Ageing: Embracing Longevity*. Barclays Equity Research, Nov. 2020. Penrose, D. *Hermes Germany focuses on next-day delivery business*.Transport Intelligence [Online]. 1 Dec. 2020. [Consulted on 5 March 2021]; via https://www.ti-insight.com/hermes-germany-focuses-on-next-day-delivery-business/

Peters, A. *These buildings combine affordable housing and vertical farming*. 18 Feb. 2021 [Consulted on 22/02/2021] via https://www.fastcompany.com/90603950/these-buildings-combine-affordable-housing-and-vertical-farming

Peters, A. *At the first lab-grown meat restaurant, you can eat a 'cultured chicken' sandwich*. 5 Nov. 2020 [Consulted on 23/02/2021] via https://www.fastcompany.com/90572093/at-the-first-lab-grown-meat-restaurant-you-can-eat-a-cultured-chicken-sandwich

Peters, A. *What if USPS delivered local produce in the mail?* 1 Mar. 2021 [Consulted on 04/03/2021] via https://www.fastcompany.com/90608098/what-if-usps-delivered-local-produce-in-the-mail

Ploegmakers, M. *Insects in feed: Sector ready to upscale*. 15 Jun. 2020 [Consulted on 26/02/2021] via https://www.allaboutfeed.net/all-about/new-proteins/insects-in-feed-sector-ready-to-upscale/

PMA (2020), *PMA's Glimpse into the Future*: 2021 Produce and Floral Environmental Scan.

Population Matters, *Welcome to the Anthropocene*, [Online]. [Consulted on 5 March 2021]; via https://populationmatters.org/campaigns/anthropocene

Rampl, L. e.a. (2012), *Consumer trust in food retailers: conceptual framework and empirical evidence*. International Journal of Retail & Distribution Management, 40 (4), 254-272.

RetailDetail. *Albert Heijn neemt 'coronaboete' op in nieuwe leveringsvoorwaarden* [Online]. 23 Feb. 2019 [Consulted on 5 March 2021]; via https://www.retaildetail.be/nl/news/food/albert-heijn-neemt-%E2%80%98coronaboete%E2%80%99-op-nieuwe-leveringsvoorwaarden

RetailDetail. *Becel-moeder brengt vegan kaas en immuunmargarine naar België* [Online]. 4 Feb. 2021 [Consulted on 16 April 2020]; via https://www.retaildetail.be/nl/news/food/becel-moeder-brengt-vegan-kaas-en-immuunmargarine-naar-belgi%C3%AB

RetailDetail. *HelloFresh rewards those who cancel* [Online]. 23 Dec. 2020 [Consulted on 5 March 2021]; via https://www.retaildetail.be/nl/news/food/beloning-voor-wie-annuleert-bij-hellofresh

RetailDetail. *Danone CEO: "Middle class about to disappear"* [Online]. 6 Aug. 2020 [Consulted on 5 March 2021]; via https://www.retaildetail.be/nl/news/food/danone-%E2%80%9Cmiddenklasse-staat-te-verdwijnen%E2%80%9D

RetailDetail. *Amazon Fresh opens up second till-free London store* [Online]. 16 Mar. 2021 [Consulted on 18 March 2021]; via https://www.retaildetail.be/nl/news/food/amazon-fresh-opent-twee-de-kassaloze-winkel-londen

Retail Detail. *The supermarket's role within the 'new normal'* [Online]. 14 Jan. 2021 [Consulted on 18 March 2021]; via https://www.retaildetail.be/nl/news/food/de-rol-van-de-supermarkt-het-%E2%80%98nieuwe-normaal%E2%80%99

RetailDetail. *Digitisation must move faster for "food tech company" Albert Heijn* [Online]. 16 Nov. 2020 [Consulted on 18 March 2021]; via https://www.retaildetail.be/nl/news/food/digitalisering-moet-sneller-voor-%E2%80%9Cfoodtechbedrijf%E2%80%9D-albert-heijn

Retail Detail. *'Duurzaamheid verkoopt niet!' (Philippe Weiler, Lidl)* [Online]. 6 Oct. 2020 [Consulted on 5 March 2021]; via https://www.retaildetail.be/nl/news/food/%E2%80%9Cduurzaamheid-verkoopt-niet%E2%80%9D-philippe-weiler-lidl

RetailDetail. *Nestlé suffers defeat, must rename 'Incredible Burger'* [Online]. 2 Jun. 2020 [Consulted on 16 April 2020]; via https://www.retaildetail.be/nl/news/food/nestl%C3%A9-moet-%E2%80%98incredible-burger%E2%80%99-hernoemen

RetailDetail. *Why Danone is becoming a flexitarian company.* 27 Aug 2020 [Consulted on 16/02/2021] via https://www.retaildetail.be/nl/news/food/waarom-danone-een-flexitarisch-bedrijf-wordt

Rijlaar, B. *De Nederlander moest Albert Heijn vertellen waar Albert Heijn voor staat.* NRC Handelsblad, 12 Jul. 2017. [Consulted on 5 March 2021]; via https://www.nrc.nl/nieuws/2017/07/12/de-pijnlijke-lessen-van-veldonderzoek-11953194-a1565825

Roque B.M., Venegas M., Kinley R.D., de Nys R., Duarte T.L., Yang X., et al. (2021) *Red seaweed (Asparagopsis taxiformis) supplementation reduces enteric methane by over 80 percent in beef steers.* PLoS ONE, 16(3)

Rüdiger Smith, T. & Yamakawa, N. *Asia's Generation Z comes of age* [Online]. 17 Mar. 2020 [Consulted on 16 April 2020]; via https://www.mckinsey.com/industries/retail/our-insights/asias-generation-z-comes-of-age

Rützler, H. & Reiter, W. (2020) *Food Report 2021.* Frankfurt: Zukunftsinstitut.

Sinke, P. & Odegard, I. *LCA of cultivated meat - Future projections for different scenarios.* Delft, CE Delft, February 2021.

Smit, A. *Powerfood for seniors.* Wageningen World 1, p. 10-15, Jan. 2015.

Snoeck, J. *Opinie: Blockchain, de killer van de platformen* 11 Jun 2018 [Geraadpleegd op 17/03/2021] via https://www.retaildetail.be/nl/news/algemeen/opinie-blockchain-de-killer-van-de-platformen

Snoeck, J. *Opinie: Blockchain, de killer van de platformen.* 11 Jun. 2018 [Consulted on 17/03/2021] via https://www.retaildetail.be/nl/news/algemeen/opinie-blockchain-de-killer-van-de-platformen

Snoeck, J. *Opinie: 'Help, er zit een Chinees in mijn broek!'* [Online]. 8 May 2018 [Consulted on 18 March 2021]; via https://www.retaildetail.be/nl/news/algemeen/opinie-%E2%80%9Chelp-er-zit-een-chinees-mijn-broek%E2%80%9D

Steel, T. *'We gaan vlees maken zoals we bier brouwen'.* 8 Oct. 2016 [Consulted on 21/02/2021] via https://www.tijd.be/ondernemen/technologie/We_gaan_vlees_maken_zoals_we_bier_brouwen/9818004.html

Stratistics Market Research Consulting (2020). *Online Food Delivery Services - Global Market Outlook (2018-2027).* Research and Markets. [Consulted on 5 March 2021]; via https://www.researchandmarkets.com/reports/5050695/online-food-delivery-services-global-market

Subramanian, S. *Is fair trade finished?* 23 Jul. 2019 [Consulted on 05/04/2020] via https://www.theguardian.com/business/2019/jul/23/fairtrade-ethical-certification-supermarkets-sainsburys

Svoboda, E. (2020). Could the gut microbiome be linked to autism? *Nature*, 577, 14-15.

Thaler, R.H. & Sunstein, C.S. *Nudge. Improving Decisions About Health, Wealth and Happiness.* Penguin, 2009.

The Business Research Company (2021). *Convenience, Mom and Pop Stores Global Market Report 2021: COVID-19 Impact and Recovery to 2030.* London: Research and Markets.

Tobin, M. & Matsakis, L. *The cut-throat war to dominate China's grocery delivery industry* [Online]. Rest of World, 20 Jan. 2021. [Consulted on 5 March 2021]; via https://restofworld.org/2021/chinas-grocery-delivery-war/

Toch, H. & Maes, A. (2020) *Scenario Agility. How to emerge stronger from the crisis.* Kalmthout: Pelckmans Pro.

Treadgold, A. & Reynolds, J. (2020). *Navigating the New Retail Landscape. A Guide for Business Leaders. 2nd Edition.* Oxford: OUP Oxford.

Ulrik, T. ed. (2020). *Fremtidens Retail 2030 I Et Udlejnings-perspektiv.* Copenhagen: Gangsted.

UNCTAD. *Maximizing sustainable agri-food supply chain opportunities to redress COVID-19 in developing countries.* United Nations Conference on Trade and Development, 2020.

United Nations Environment Programme (UNEP). *Food Waste Index Report 2021.* Nairobi, 2021.

United Nations Environment Programme (UNEP). *Worldwide food waste* [Online]. [Consulted on 5 March 2021]; via https://www.unep.org/thinkeatsave/get-informed/worldwide-food-waste

Vandist, S. *Pretopia.* Lannoo Campus, 2021

Van De Velden, W., *Stikstof dreigt Vlaamse economie te verstikken.* De Tijd, 3 Mar 2021. [Consulted on 5 March 2021]; via https://www.tijd.be/politiek-economie/belgie/vlaanderen/stikstof-dreigt-vlaamse-economie-te-verstikken/10288420.html

Van der Weele, C. et al. *Meat alternatives; an integrative comparison.* Apr. 2019 [Consulted on 28/02/2021] via https://www.researchgate.net/publication/332717011_Meat_alternatives_an_integrative_comparison

Van Geyte, J. *Denmark has supermarket for excess food.* 24 Feb. 2016 [Consulted on 09/03/2021] via https://www.retaildetail.be/nl/news/food/denemarken-heeft-supermarkt-voor-voedseloverschotten

Van Huis, A. et al. *Edible insects: future prospects for food and feed security.* 16 Aug. 2013 [Consulted on 24/02/2021] via http://www.fao.org/3/i3253e/i3253e.pdf

Van Rompaey, S. *Delhaize opent stadsmoestuin.* 18 Oct. 2017 [Consulted on 05/02/2021] via https://www.retaildetail.be/nl/news/food/delhaize-opent-stadsmoestuin

Van Rompaey, S. *Colruyt start hoogtechnologische kruidenteelt.* 3 Mar. 2020 [Consulted on 08/02/2021] via https://www.retaildetail.be/nl/news/food/colruyt-start-hoogtechnologische-kruidenteelt

Van Rompaey, S. *"European retailers can learn from Walmart" (Peter Hinssen, nexxworks)* [Online]. 4 Apr. 2019 [Consulted on 18 March 2021]; via https://www.retaildetail.be/nl/news/algemeen/%E2%80%9Ceuropese-retailers-kunnen-wat-van-walmart-leren%E2%80%9D-peter-hinssen-nexxworks

Van Rompaey, S. *The fight over vegan consumers turns into a war.* 4 Jan. 2019 [Consulted on 16/02/2021] via https://www.retaildetail.eu/en/news/food/gevecht-om-de-vegan-consument-wordt-titanenstrijd

Various. *I can't believe it's not meat.* 22 May 2019, Barclays Equity Research.

Vergeer, R., Sinke, P., and Odegard, I. *TEA of cultivated meat. Future projections of different scenarios.* Delft, CE Delft, February 2021.

Vincent, A., Stanley, A. and Ring, J. *Hidden champion of the ocean – Seaweed as a growth engine for a sustainable European future.* 2020 [Consulted on 27/02/2021] via https://www.seaweedeurope.com/wp-content/uploads/2020/10/Seaweed_for_Europe-Hidden_Champion_of_the_ocean-Report.pdf

Vilt, *Crevits wil lokale voeding op de kaart zetten.* 3 Sep. 2020 [Online]. [Consulted on 5 March 2021]; via https://vilt.be/nl/nieuws/crevits-wil-lokale-voeding-op-de-kaart-zetten

Wepner, B. et al. *FIT4FOOD2030 – Towards FOOD 2030 – future-proofing the European food systems through Research & Innovation.* Aug. 2018, AIT (Austrian Institute of Technology).

Wepner, B. et al. (2021) *Attachment 6.5 to Deliverable 2.1. Report on baseline and description of identified trends, drivers and barriers of EU food system and R&I - Description of Trends.* FIT4FOOD2030 [Online]. [Consulted on 16 April 2020]; via https://fit4food2030.eu/reports-publications/

Willett W., Rockström J., Loken B. et al. *Food in the Anthropocene: the EAT–Lancet Commission on healthy diets from sustainable food systems.* 16 Jan. 2019 [Consulted on 10/02/2021] via https://eatforum.org/eat-lancet-commission/eat-lancet-commission-summary-report/

Witte, B. et al. *Food for Thought - The Protein Transformation.* Boston Consulting Group & Blue Horizon Corporation, 2021.